THE
VERTICAL ASCENT

From Particles
to the Tripartite Cosmos
and Beyond

THE
VERTICAL ASCENT

From Particles
to the Tripartite Cosmos
and Beyond

Wolfgang Smith

Philos-Sophia Initiative Foundation

To request permission, contact the publisher at
info@philossophiainitiative.com

Hardcover ISBN: 978-1-7359677-0-7
Paperback ISBN: 978-1-7359677-1-4
eBook ISBN: 978-1-7359677-2-1

Library of Congress Control Number: 2020922071

Philos-Sophia Initiative Foundation
www.philos-sophia.org

To Rick DeLano
in high esteem and eternal gratitude

TABLE OF CONTENTS

	Foreword	ix
	Prologue	xv
	Preface	xxi
1.	To Be or Not To Be an Apple	1
2.	The Tripartite Wholeness	13
3.	From Cosmos to Multiverse: The Ominous Descent	27
4.	Lost in Math: The Particle Physics Quandary	43
5.	Do We Perceive the Corporeal World?	59
6.	Pondering Bohmian Mechanics	71
7.	Astrology: The Science of Wholeness	85
8.	Evolutionist Scientism: Darwinist, Theistic, and Einsteinian	99
9.	Vertical Causation and Wholeness	111
10.	The Mystery of Visual Perception	121
11.	Gnosticism Today	137
12.	Beyond the Tripartite Cosmos	155
13.	Does Physics Admit a Teleology?	169
14.	Science, Scientism, and Spirituality	173
	Index	181

FOREWORD

Olavo de Carvalho

THERE HAVE BEEN countless studies on the evolution of knowledge, but none, or next to none, as far as I know, on the evolution of ignorance. However, a cursory examination is enough to reveal that the knowledge we have lost is just as vast and valuable as the body we have acquired. The men who arranged the Stonehenge monoliths in a circle knew exactly what they were doing and why, though we have yet to find a satisfactory explanation for it ourselves. We have a thousand unproven theories as to what the pyramids of Egypt were for and how they were built, but the more we read about the science of the Pharaohs in the monumental *Le Temple de l'Homme*, by the Polish archaeologist Schwaller de Lubicz, the more we have to confess that, all things considered, we understand nothing. *Homo neanderthalensis* knew on clear and present evidence whether it was anthropoid or man, but we're still banging our heads together over it. And the more we laud and applaud the authority of science, the more soaringly abundant cases of scientific fraud chip away at the very credibility of science itself. Indeed, the pathetic showing the World Health Organization has made of itself throughout the COVID-19 pandemic has transformed the very idea of a global health authority into the butt of equally worldwide ridicule.

Worse. There is no certainty whatsoever that the much vaunted progress in knowledge amounts to anything but a metastasis in material records all but unserviceable by even the most superior human intelligence. I sometimes ask myself whether our entire corpus on mineralogy stowed away in libraries, archives, on microfiche and video comes to anything but a compilation of doubts and questions matched for volume only by the vast horde of minerals we have yet to study at all. Archival science, by electronic means or any other, has become such a broad and unencompassable branch of knowledge that a whole lifetime of study can afford no certainty of ever mastering it. Why are there no scientists honest enough to admit that even the indexes of their chosen fields of expertise, much less the contents of those fields, lie infinitely beyond the scope of their vision? And what credence are we supposed to give to professionals who appear so woefully unaware of even the most obvious limits of their own capacities?

All the evidence suggests that "progress of knowledge," no matter how firmly people believe in it, does not correspond to any palpable, describable reality: it is a metonym, the name given to a profession of faith, or, better put, of a problem—an unsolvable problem, thus far, at any rate. In fact, we cannot measure what we have learned if we do not know how much we have forgotten. Without the history of ignorance, belief in the progress of knowledge is just a universal case of the Dunning-Kruger effect.

But the most serious case presents itself, not when knowledge is lost, but when the academic elite and general accepted opinion come to the consensus—an unassailable certainty endowed with all the gravity of reality—that it never existed anyway.

The present book by Dr. Wolfgang Smith is about such a case. The author has broached the issue in earlier works, but here he picks it apart in such dramatic detail that it becomes impossible for the honest reader to fail to realize that the episode in question is no mere mistake, but an epistemological scandal of colossal, tragic, and intolerable proportions.

Dr. Smith calls this phenomenon bifurcation. It refers to the ostensibly irreducible distinction Descartes, Newton, and other scientists on the threshold of the Modern Age established between the primary and secondary qualities of material objects. The former are mensurable characteristics—extension and weight, for example—and constitute the only objective reality of things. The latter, such as color or smell, are merely alterations occurring within the body and mind of the human observer, and which we can never be certain actually exist in the external world.

The same idea that mensurable qualities exist in the perceived object itself and not in the mind that measures them is already odd to start with, seeing as (1) the terms "measure" and "measuring" come from the Latin *mens*, which means mind; and (2) all measuring is a comparison between two objects, or between an object and a given standard, such that it is inconceivable that a standalone object free-floating in space could have any "measurements" in and of itself.

Secondly, based on such studies as James Gibson's *The Ecological Approach to Visual Perception* (1986), Prof. Smith correctly stresses that if anything is absolutely impossible it's that the sensible qualities perceived in objects exist "in our minds," or anywhere else in our bodies (or "souls"), and not in the things themselves. In fact, if these qualities exist in the brain only, their measurements would be there too, necessarily. The world of so-called primary qualities would therefore be every bit as subjective as the secondary.

Despite or even because of this intrinsic absurdity, bifurcation served as a pretext for the most ambitious grab for authority the world has seen since those festive days when Julius Caesar declared himself the direct descendant of the goddess Venus. If only entities measured by physicists are objectively real and everything else exists only in the fantastical mind of an unversed humanity, the conclusion is unavoidable: physicists alone can know reality and distinguish it from fantasy. Otherwise put: either you obey the physicists, or you are a total nut-job. Of course, no physicist would be nutty enough to declare this publicly, but amongst themselves, and deep inside, many of them fervently believe it to be true. Just wait for the day they come out proclaiming themselves descendants of Venus, too.

Such is their belief that "scientific consensus" has become the last word not only on science-related subjects, but on philosophical, moral, religious, political and psychological issues too, and yet they fail to realize that a "consensus" is just a tally of votes endowed with no deeper objective reliability than any rhetorical argument.

How, we might ask, did Descartes, the prince of modern rationality, manage to spawn so much claptrap? From the outset, I don't believe Descartes was as rational as he's made out. In his most important work, *Meditations on First Philosophy*, he promises to recount cognitive experience succeeding upon his personal, historical "I," but then, all of a sudden and apparently without noticing it, he leaps from that narrative of concrete psychological facts to the pure analysis of a generic, abstract philosophical "I," unwavering in his belief that he is still talking about his real, biographical self.

In my booklet "Visions of Descartes," I showed that the experience of radical doubt as Descartes formulates it is a total psychological impossibility. When a man says he has done something we realize to be impossible, we have to conclude that he either did nothing at all or something else entirely, which he has since misnamed. If Descartes can't have pulled off the experiment he says he did, then what, if anything, did he do?

I can't think of a single Descartes scholar who has noted the paroxysm of inattention with which the philosopher springs from his biographical "I" to this abstract concept of the philosophical "I," but likewise none seems to have even raised the hypothesis that the famous "evil genius" might be more than just a literary artifice created to facilitate the exposition of this abstract idea, but the actual account of a perfectly real inner experience, an authentic religious obsession. Clamoring loud and clear in favor of this hypothesis is the fact that, having proclaimed the *ego cogitans* as the source of all certainty, the philosopher proceeds to recognize that this ego grounds no certainty whatsoever beyond its own existence, a subjective prison from which it can only be delivered by... appealing to God. So rather than a pure theoretical question of epistemology, what we've really got is a theological drama resolved by theological means. Resolved, that is, by theological means dressed up as theoretical epistemology—hardly surprising for a man whose life maxim was "Masked I go forward."[1]

If, over the centuries, the Cartesian schism, affirming itself the sole absolute certainty and accepted as such by much of academia, churned out so many absurdities—and this book will evince some of the worst of these—it may have a lot to do with the disguises and deceptions that lie at its very origin.

---

1. Those who have not read Maxime Leroy's masterly biography of Descartes, *Philosophy au Masque* (Paris: Éditions Rieder, 1929), will have no idea of the extent to which pretense and camouflage were decisive factors in the philosopher's life and work.

A late friend of mine, a genius of clinical psychology, once told me: "Neurosis is a forgotten lie in which you still believe." The experience of life teaches us that when things get inextricably muddled it's usually because some forgotten falsehood is still at work behind reams of shadow and layers of camouflage.

The bifurcation is wider, more active, and more lethal than these falsehoods. Dr. Smith has discussed it in earlier works, but in the present book he goes beyond critical examination to develop an alternative theory—a "tripartite wholeness"—with which to free science from this four centuries-old neurosis.

I do not believe the current generation of scientists and intellectuals can boast many among their ranks who are capable of gauging the true measure of the scientific overhaul Prof. Wolfgang Smith has performed in his oeuvre, but in this work he takes it to its undeniable conclusion. However, it is inevitable that the coming centuries will recognize that what Prof. Smith has done with science was more than a reform; more, indeed, than a psychoanalytical cure. It was, first and foremost, an exorcism.

PROLOGUE

EVEN AS THERE ARE, according to Wolfgang Smith, "two apples"—the *physical* and the *corporeal*—there are likewise in principle "two physicists": the one subject to the confusion of scientistic belief, the other astute enough not to be thus deceived. The first a victim of Cartesian bifurcation—the notion that the external world reduces to sheer quantity in the form of so-called *res extensae*, while all the rest pertains to a subjective realm comprised of *res cogitantes* or "things of the mind"—the other a physicist who at the same time is also a competent metaphysician. It is worth noting that whereas in days gone by it was not that unusual to encounter a metaphysician who was also a physicist, the reverse is rarely to be found. "To be or not to be a metaphysician": this then is the crucial issue.

Now such a twofold comprehension of the cosmos—*qua* both scientist and metaphysician—constitutes in fact the salient characteristic of Wolfgang Smith. Drawing upon a profound grasp of foundational physics—along with certain other basic scientific disciplines, such as neuroscience and astrophysics—it is his wont to penetrate into the underlying metaphysics, thereby casting the issues in a brand new light. Typically it turns out that, having done so, the *prima facie* difficulties that motivated the inquiry—which appeared insurmountable on the level at which they were originally conceived—prove to be readily tractable when thus viewed on deeper ground.

The reason resides in the fact that the problems in question—though seemingly scientific—are in truth philosophical, which is to say that they arise from "the bad metaphysics" of the physicists. And this is something Wolfgang Smith has dealt with extensively, what he terms "scientism" as distinguished from science properly so called. Scientism is in truth an insidious *Weltanschauung* masquerading as Science, which tends to mislead just about everyone by virtue of the fact that it epitomizes the prevailing *Zeitgeist*, the very outlook definitive of our times. And it is from this generally unrecognized and unsurmised plague of scientistic belief that Wolfgang Smith offers relief, rescue and deliverance. Penetrating the veneer of "scientific" respectability, he takes us beyond the fantasies and unconscious assumptions definitive of the scientistic credo.

It is to be emphasized that this constitutes in no wise a critique or infringement upon science as such, but an admonition, rather, to remain true to its fundamental principles. What needs to be unmasked and renounced is not science, but scientism, which is not only something utterly different, but is itself at odds with the *modus operandi* of science as such.

To be precise, science is a "knowledge through causes"; as the Scholastics say: *scientia est cognitio per causas*. But whereas *philosophy* is a knowledge through first and universal causes, science (in the contemporary sense) is a *cognitio* from secondary causes. Wolfgang Smith's distinction between "vertical" and "horizontal" causation—a causality which is instantaneous and unmediated, as distinguished from a causation transmitted through space by some temporal process, is a case in point.[1]

Science as we understand the term entails moreover a double reduction based upon the distinction between the *material object* (inorganic, plant, or animal for instance) and its *formal object* (physics, chemistry, astronomy say), the point being that a science is defined not simply by the former but depends vitally on its formal object as well. Given, for example, that plants constitute the material object of both botany and pharmacology, it is their formal object that distinguishes these disparate sciences. The point is that *what* we know is a function of *how*—by what means—we know it. And this explains why there are in truth "two apples": the *corporeal* and the *physical*, as Wolfgang Smith insists.

The danger with science *per se* resides in the fact that it may transgress—or claim to transgress, better said—the bounds imposed by the aforesaid double reduction. When physicists, for example, allude to a "theory of everything," they have clearly crossed the line. The now "classic" example of such an overreach

---

1. See *Physics and Vertical Causation* (Angelico Press, 2019).

appears to be Stephen Hawking's bestselling treatise *The Grand Design*, which Wolfgang Smith has unmasked as the epitome of scientistic transgression.[2]

~

The validity of science—its veracity, one can say—hinges upon the recognition of the aforesaid double reduction upon which it is based. Whereas Plato gave expression to metaphysics—to what it is and must be—Aristotle laid the foundations of science as such. These foundations were however jettisoned in the latter half of the second millennium, beginning with the Galilean subjectification of the qualities: the hegemony of what René Guénon refers to as "the reign of quantity" had thus begun. As Jean Borella points out, a radical and ruthless "geometrization" of the cosmos had commenced, which in fact attained its theoretic completion in the Einsteinian reduction of time to space by way of a 4-dimensional "space-time," which as Wolfgang Smith points out, proves to be "a bridge too far."[3] When it comes to foundational—or so-called "particle"—physics in particular, the scientific method based on empirical verification has in effect been abandoned in favor of a kind of mathematical universe building, which for about the past half century has lived "on credit" so to speak, secured by the spectacular achievements dating back to the "glory days" following the discovery of quantum mechanics as conceived by the likes of Niels Bohr and Werner Heisenberg. As Wolfgang Smith enables us to recognize, we are presently nearing the end of that "reign of quantity" initiated four centuries ago by the Galilean reduction: as the title of the Philos-Sophia film has it, "*The End of Quantum Reality*" appears to be at hand.

What is urgently needed at this critical juncture is a smattering, at least, of authentic metaphysics, enough to unmask and dispel the Galilean-Cartesian bifurcationism. This spurious tenet

---

2. See *From Physics to Science Fiction: Response to Stephen Hawking* in *Science & Myth* (Angelico Press, 2010), ch. 7.

3. See *Physics and Vertical Causation*, op. cit., ch. 5.

has become ingrained in the scientific mentality of our day to the point of being *de facto* invisible, and hence unassailable as well.

~

It should however be pointed out—in the name of historical accuracy—that actually René Descartes was not a Cartesian, nor was Sir Isaac Newton a Newtonian. To begin with the so-called Cartesian "mind-body" dualism, what Whitehead terms "bifurcation": the distinction between *res cogitans* and *res extensa* was more methodological than ontological. For René Descartes, the one and only actual entity is the *individual*—that is to say, quite literally, the *indivisible*. As he tells us explicitly in the *Meditations*: "I am intermingled with it [i.e., the corporeal *res extensa*], so that I and my body form a single entity." Thus, for Descartes, the junction of body and soul constitutes a *substantial union*. This means that the famed "Cartesian mind-body dualism" is *de facto* an invention on the part of historians rather than an historical fact.[4]

So too it happens that Sir Isaac Newton was not in truth a so-called Newtonian. For him the force of gravity, for instance, exemplified the shortcoming of physical science as such: a manifestation, namely, of superior forces manifesting God's presence and action within the universe. He thus perceived it in fact as an instance of what Wolfgang Smith refers to as "vertical causation," which acts instantaneously and does not entail a transmission of some kind through space. It was only after his dispute with the supposed "etherists" that Newton was led to abandon public reference to metaphysical or theological conceptions in favor of a "methodological positivism" of a kind, in deference to his opponents. This turned out moreover to be a Pyrrhic victory for the latter, as Alexandre Koyré points out. Meanwhile Newton persisted in offering metaphysical as well as theological interpretations of what we deem to be purely physical phenomena, based on his conception that absolute space—so far from being simply an empty

---

4. On this issue we refer especially to the studies by André Charrak. See also our introduction to Jean Borella & Wolfgang Smith, *Rediscovering the Integral Cosmos: Physics, Metaphysics, and Vertical Causality* (Angelico Press, 2018).

receptacle—constitutes what he termed the *sensorium Dei*: the modality, that is, through which God is present to all things. The Newtonian conception of space proves thus to be the very opposite of what we refer to as "Newtonian."

Yet regardless of whether Descartes or Newton are responsible or not, the twin doctrines promulgated in their name are in truth flawed, as Wolfgang Smith has shown, to the point of rendering a correct understanding of physics impossible.

On the other hand—guilty doubtless, *a contrario*—are Galileo and Kant, whose respective doctrines have plunged the Occidental world into a state of *de facto* metaphysical insanity. As regards Galileo: genius though he was—recall his "law of motion"—his vision of a universe comprised of raw "matter" moving according to mathematical laws turns out to be a half-truth at best. Kantism, on the other hand, may well be the most dangerous doctrine of all. For whereas reason is in truth subordinate to intellect, properly so called, Kant inverts the hierarchic order by proclaiming reason to be supreme, while reducing it to the psychological realm. The result of this inversion—amply corroborated by the subsequent history of Occidental philosophy—has been to render the external world absolutely unknowable and thereby reduce metaphysics to an illusion. Caught between the options, on the one hand, of an external world reduced to raw matter, and on the other a realm which must forever remain unknown, the physicist found himself from the start in an unpromising position.

This is where Wolfgang Smith enters upon the scene. What he offers is, in substance, the one and only way out of the Galilean-Kantian impasse: a return, namely, to sound ontology, which he discerns in the great metaphysicians of antiquity, from Plato and Aristotle to St. Thomas Aquinas. What he effects is a return to metaphysics freed from the stultifying axioms of Galileo and Kant, thereby rescuing physics from its own demons.

∿

No need, then, to belabor this culminating opus of Wolfgang Smith—which is both a summation and completion of his earlier work—with additional commentary. Suffice it to say that whereas in this truly magisterial treatise he revisits "peaks" previously ascended, *The Vertical Ascent* takes us to heights not previously attained. For whereas the journey "from particles to the tripartite cosmos" has in fact been summarized in his recent book *Physics and Vertical Causation*, his *Vertical Ascent* refers to a *Beyond* exceeding even the reach of metaphysics *per se*. There is in fact no room anywhere in this new book for a *déjà-vu*: *The Vertical Ascent* is charged with its own sense of discovery, and at every turn brings into view vistas not previously revealed.

What enables this breakthrough, to be sure, are the fundamental clarifications achieved in his earlier publications. Key metaphysical conceptions had already been put in place and properly linked: whether it be the notion of a wholeness "greater than the sum of its parts," the tripartite "*corpus, anima, spiritus*" anthropology—not to speak of the key notion of *vertical causation* at which Smith arrived through his resolution of the measurement problem in quantum mechanics—all these enter crucially into *The Vertical Ascent*, howbeit from an enlarged point of vantage. There are also, moreover, numerous topics—from "astrology" to "Gnosticism"—never previously broached, which now enter crucially into play. Yet the fact remains that *The Vertical Ascent* constitutes in essence a single "theorem": a single work of art one can say, in which not a brushstroke can be altered without detriment. Most important of all, however, this *magnum opus* stops not with the rediscovery of the tripartite and thus inherently Platonist cosmos, but goes on to take the ultimate step in pointing to that ineffable *Beyond* which proves to be the authentic Eschaton of mankind.

"To Be or Not to Be a Metaphysician": that is indeed the Question, the one that counts in the end. May this book then—this *magnum opus* of Wolfgang Smith—bring out the true Metaphysician in us all: the helmsman to guide us to the threshold of Eternity.

Bruno Bérard

PREFACE

THIS BOOK WAS not planned, but consists of fourteen articles written for the website of the Philos-Sophia Initiative Foundation, which add up to form a whole: a vision namely of the hierarchic cosmos. That vision is founded upon what I term the "cosmic icon," which outwardly is simply a circle, comprised of a center, an interior, and a circumference. What renders that figure ontologically meaningful and indeed iconic, is that we associate two of its three components with the cosmic bounds of time and space: the interior, namely, with time alone, and the circumference—representing our corporeal world—with both. Thus iconically conceived, the integral cosmos proves to be ontologically tripartite, consisting in ascending order of a corporeal domain subject to space and time, an intermediary realm subject to time alone, and finally a Center subject to neither bound. I surmise that this symbolic representation of the integral cosmos was in some way recognized in the major premodern schools, and am in any case persuaded that it provides the key to an ontological understanding of what has sometimes been referred to as the *cosmologia perennis*.

It hardly needs pointing out that the subtitle "From Particles to the Tripartite Cosmos and Beyond" alludes to the three domains in question, and refers to an ontological ascent. Now the first step in this "vertical ascent" consists in the transition from the realm of quantum particles—what I term the *physical* domain—to the corporeal. The ontology of this transition emerges from a consideration of the so-called measurement problem for quantum mechanics, which demands a transition from the physical to the corporeal domain, as I shall argue in Chapter 1.

Initially the measurement problem was understood to center upon the question whether the so-called collapse of the wave function in the act of measurement could be explained in quantum mechanical terms. The influential Copenhagen school, beginning with Bohr and Heisenberg, claimed it could not—that the measuring instrument could not be conceived as being itself a quantum system. The measurement problem can thus be seen as a challenge to circumvent the Copenhagenist "dualism" of quantum system plus measuring instrument, and for almost a hundred years top-ranking physicists attempted to accomplish this feat. Some—like David Bohm, for instance—have claimed success. For my part, I viewed the matter from an ontological point of vantage which enables one to conclude quite readily that the measurement of a quantum system demands a *corporeal* instrument: the *measurable* cannot measure itself. Meanwhile a small but distinguished group of physicists have in recent times sided with the Copenhagenists to the effect that the measuring instrument cannot be itself conceived as a quantum system, and have devised theories of measurement in which the instrument is successfully treated more or less as an object of classical physics.[1]

How then does the second step come about: the ascent namely from the corporeal to the intermediary? We approach this question by way of the famous "binding problem" in the neuronal theory of visual perception: we propose to show that this problem cannot in fact be resolved on the corporeal plane—which explains why, after more than half a century of concentrated research, the problem looms larger than it did at the start. We shall argue that the binding in visual perception actually takes place on the intermediary level, by virtue of the fact that on this plane spatial separation as such is transcended.

This brings us to the third and final ontological leap: from the intermediary to that supreme time-transcending realm, what Plato refers to as the "intelligible" world: what is it then that necessitates an ontological transition of that kind? We tackle the issue in the context of visual perception, based in part on James J. Gibson's "ecological" theory, by establishing that the perception of motion calls for a supra-temporal point of vantage. —These then, in all

---

1. I would point out that whereas I incline to regard something like George Ellis's theory of "Contextual Wavefunction Collapse" as a highly respectable physical theory, I nonetheless surmise that the crux of the matter eludes its grasp by virtue of being—not physical—but ontological.

brevity, are the three ontological steps leading to a rediscovery of the "tripartite" cosmos known apparently to the ancient schools, from Plato to Aquinas—which to this day is referred to among educated Hindus as the *tribhuvāna* or "triple-world."

But whereas these three steps consummate the ascent "from particles to the tripartite cosmos," and do so on basically scientific and philosophical grounds, the question remains: what about that "Beyond" likewise alluded to in the subtitle? Here we need to shift gears: there is admittedly no scientific or philosophical argument to validate—or even *define*—that ultimate step. Yet, by the same token, neither is there any such argument *against* that possibility. Hence we are entirely unconstrained on grounds of human rationality, free to pursue the question—or leave it be. Not everyone, therefore, will follow us on that "fourth"[2] stretch of the Vertical Ascent. In the last three chapters we thus address ourselves primarily to readers prepared to approach the Old and New Testament texts with the respect they demand, open at least to the possibility of a supra-cosmic Beyond. What stands at issue is what Christians refer to as "the Kingdom of God" and regard as the ultimate destination to which they are called. It needs to be clearly understood that this Christian Eschaton transcends not only the bounds of space and time, but even that supreme cosmic realm Platonists refer to as the "intelligible" world. That Kingdom, thus, is something so radically "*not of this world*" as to be incomprehensible even to the highest philosophy: as we have said, the concept pertains—not to philosophy, however exalted—but to the domain of Revelation.

Getting back to the cosmological portion of the Ascent: along the way we touch upon various foundational matters which can only be understood from an ontological point of vantage and have now become comprehensible. In chapters 3 and 4, for example, we show that physics reduced to its ultimate form through the

---

2. Readers acquainted with the Upanishadic literature may recall that the *Mandukya Upanishad* refers to that "supra-cosmic" realm beyond the *tribhuvāna* simply as *turiya*, which means "the fourth." And this is wise: for such is in essence all that human wisdom as such can declare concerning that supreme sphere.

elimination of the non-empirical notion of "matter" *qua substance* reduces inevitably to quantum theory, whereas the Einsteinian "geometrization of time"—being ontologically invalid—cannot but lead eventually to a *reductio ad absurdum*, be it in the form of a multiverse, superdeterminism, or something else no less spurious.

Or to cite an even more basic example: the ascent from the physical to the corporeal plane in the act of measurement brings into play the first manifestation of *wholeness* in the form of a hitherto unrecognized mode of causality arising no longer from parts—as does the causality upon which physics as such is based, what I term *horizontal* causality—but originating from wholeness, what I refer to as *vertical* causation. And this fundamental recognition leads quite readily to another: while horizontal causality, by virtue of its spatio-temporal nature, is restricted to the corporeal plane, vertical causation—which springs from wholeness and involves no spatio-temporal transition—is operative without restriction, and proves therefore to be the primary mode. As such it gives rise to the horizontal, and has, moreover, the capacity to override the latter, as happens, for instance, in the act of measurement.

In the wake of this absolutely fundamental recognition, we will note that it is in fact a characteristic of the ancient or premodern sciences to be based—not, like our physics, upon a reduction to infinitesimal parts—but upon wholeness, precisely, which entails that these ancient sciences operate primarily by means of *vertical* causation. Following this discovery, we devote an entire chapter to one of the most ancient and profoundly misunderstood sciences: i.e. astrology, which proves to be the diametric opposite of physics. For whereas the latter is based upon a radical decomposition of wholes into their smallest component parts, astrology operates purely and simply with wholes, and consequently with vertical causation. It thus turns out that the shoe is now on the other foot: when we denigrate such ancient sciences as "pre-scientific superstitions," the onus of ignorance is on us.

Let me not neglect to point out another major feature of the book: it picks up, namely, and further develops the categorical distinction between *science*, properly so called, and *scientism*, which is "science" in name only, and reduces, in fact, to an *ideology*. As readers acquainted with my work will know, I have been harping on that subject ever since the first chapter of my first book, in which I identified many prestigious so-called "scientific" teachings—from Darwinist evolution to Freudian and Jungian psychology—as so many varieties of vintage scientism. In the present book I take this inquiry a decisive step further by showing that contemporary scientism at large proves in fact to be a new form of that ancient counter-religion known to scholars as *Gnosticism*. To be precise, I show that the classical Gnosticism, associated with such figures as Simon Magus and Valentinus, has morphed in post-medieval times into what may properly be termed *neo-Gnosticism*, which so far from constituting some outlandish cult, is in fact the veritable heart and center of contemporary so-called "progressivism" in all its diverse forms. Its "central myth" proves to be none other than the doctrine of a universal and all-encompassing Evolution, which shall supposedly empower humanity to attain its true Eschaton, now conceived in futuristic and collective terms. Catholics, in particular, may be surprised to recognize Pierre Teilhard de Chardin as the neo-Gnostic guru *par excellence*, its veritable "prophet" no less, who under the guise of a "renewed Christianity" preached the gospel of that neo-Gnostic Eschaton to the rapturous acclamation of millions! And by way of contrast, now, let us recall the authentic Christian Eschaton, which—so far from constituting the final stage of history—resides, as we have noted, beyond the "intelligible world" of Platonist philosophy, transcending even its aeviternity.

Having contemplated the Vertical Ascent culminating in the recognition of the supreme Eschaton, the question arises whether that comprehension can in turn tell us something concerning the tripartite cosmos which otherwise could not be understood. This proves to be in fact the case, and we shall offer an example in chapter 13. Based on Christian sources, we establish a fundamental fact regarding physics, which physics as such cannot so much as conceive: the fact, namely, that *physics admits a teleology*. To be precise, we show that the so-called Principle of Least Action—from which the laws of motion in both classical and quantum physics can be deduced—admits of a teleological interpretation based upon Christian metaphysical teaching.

The final chapter is addressed, above all, to Catholics perplexed by what has befallen the Church since Vatican II. It argues that the scientistic *Weltanschauung*—or more accurately, the imposition of that *Weltanschauung* upon the public at large, inclusive of priests, religious and theologians—has imperiled its very survival on a human plane. As Jean Borella has put it:

> The truth is that the Catholic Church has been confronted by the most formidable problem a religion can encounter: the scientistic disappearance (*disparition scientifique*) of the universe of symbolic forms which enable it to express and manifest itself, that is to say, which permit it to exist.

Admitting that other factors are likewise in play—that, above all, "*we wrestle not against flesh and blood, but against principalities, against powers, against the rulers of the darkness of this world*"—it can even so be said that the current hegemony of scientism suffices in itself to explain the present devastation. The "*principalities*" and "*powers*" to which St. Paul alludes do no doubt play a necessary and indeed essential role; my point is that they may achieve their end simply by imposing the yoke of scientism upon humanity: for this in itself suffices to deprive the Church of that which "permits it to exist." As St. Thomas Aquinas makes clear: what we think regarding this world is ultimately crucial to the life of Christianity.

One final comment. Inasmuch as the chapters originated as independent articles, a certain bit of repetition remains in the text. This need not, however, be seen as a shortcoming. The ideas, first of all, are challenging enough to bear repetition. Moreover, these repetitions may serve to exhibit the continuity, the "wholeness" if you will, of the text in its entirety. And lastly, they render the chapters "independently comprehensible," which some readers, at least, might find helpful.

1

TO BE OR NOT
TO BE AN APPLE

A Brief Response to Dr. Alec MacAndrew

THE TITLE ALREADY of Dr. MacAndrew's article—"When Is an Apple Not an Apple?"—alludes to the precise point of contention between us in that it differentiates *my* interpretation of physics from the current: where the contemporary consensus affirms *one* "apple," I recognize *two*. Let me explain.

The contemporary and reputedly "scientific" worldview stipulates that what we actually perceive through our senses pertains to a subjective ontological domain, whereas the "real" world reduces to the *physical*: i.e., to the universe *as conceived by the physicist*. For my part—and in keeping with the sapiential traditions, from Plato and Aristotle to St. Thomas Aquinas—I flatly disagree. To claim, as one nowadays does almost automatically, that cosmic reality reduces thus to the categories of physics—to entities describable *without residue* in *mathematical* terms—that, in my opinion, is not only unproved and unprovable, but patently *absurd*. In addition to quantities or "quantitative attributes," I maintain, cosmic reality comprises perforce *qualitative* attributes as well, which in fact is precisely what renders that world *perceptible*. In a word, I place myself on pre-Cartesian, pre-Enlightenment ground in distinguishing what I term the *corporeal* domain from the *physical*: the world as "perceived"—or better said, as humanly *perceivable*—from the world *as conceived by the physicist*. Much as I respect and admire his expertise, I don't regard his wisdom to be all-encompassing. On the contrary: I surmise that the efficacy of the physicist's *modus operandi* derives actually from the limitation of his purview, the narrowness of the "slit" through which he examines the universe.[1]

This brings us to the "two apples": there is the "red apple" we can perceive, hold in our hand, and bite into: the "corporeal apple" I call it; and there is, in addition, the "physical apple," made up of protons, neutrons, and electrons, or more generally, of "quantum stuff" howsoever conceived. And these two entities—though they occupy the same spatio-temporal locus—are manifestly *not* the same! Which brings us to the very point that differentiates my ontology from that of Dr. MacAndrew and the physics community at large: whereas *they* believe in "one apple"—assume in other words that the *"corporeal"* reduces to the *"physical"*—I reject this reductionist postulate, and maintain that, on the contrary, the corporeal apple takes precedence over the physical. I claim in fact that the relation between the two can be conceived in Aristotelian terms: that *the physical*, namely, *stands to the corporeal as potency to act.*[2] To Dr. MacAndrew, on the other hand, the corporeal appears to be "Not an Apple," by which he apparently means that it does not in truth exist. His message, I take it, is that *there are in fact not two "apples," but only one*: the *physical* namely.

This, then, appears to be the basic point of contention between us, what Dr. MacAndrew—in company with the physics community at large—objects to primarily in my ontological interpretation of physics. What stands at issue thus is whether *the perceptible world* reduces to "quantum stuff" or not. And as the title of the Philos-Sophia Initiative film—*The End of Quantum Reality*—implies, I maintain that the objectively real world does *not* in truth reduce to the conceptions of the physicist.

---

1. The idea goes back at least to Goethe's dictum: *"In der Beschränkung zeigt sich der Meister"* ("In limitation the master shows himself").

2. This idea goes back actually to Werner Heisenberg, who made it a point to observe that quantum particles are not in fact real entities, but something "midway between being and nonbeing" which as such "are reminiscent of Aristotelian *potentiae.*" Of all the great physicists, he appears to have had the deepest grasp of metaphysics, and came, I believe, within a hair's breadth of rediscovering the corporeal domain.

Yet, as if this contention were not in itself aggravating enough to the physics community, I compound the offense by showing that *this very Ansatz resolves the famed "measurement problem"*—which despite prestigious contentions to the contrary, has *not* in truth been solved for the physicist as such. Neither the *tour de force* of Bohmian mechanics, nor "many-worlds" theory, nor "decoherence," nor indeed any of their proposals has yet succeeded in breaking the impasse.[3] I am in fact persuaded that there *is no resolution of the measurement problem* in the "one apple" ontology: given that the act of measurement entails a "collapse of the wave function"—which the Schrödinger equation itself cannot yield—it follows that, in a "quantum world" governed by a "cosmic" Schrödinger equation, *there can be no such collapse.* On the other hand, once we acknowledge that the measuring instrument is in fact *corporeal*—which in my view it *has to be*, on pain of being imperceptible—the difficulty vanishes, and the resolution of the measurement problem stares us in the face.

∾

Having clarified what apparently stands at issue in Dr. MacAndrew's allusion to "Apples that are not Apples," I would point out that he is "on target" in focusing his attack on the one-and-only treatise I have devoted exclusively to the explication of the offending tenet: i.e., *The Quantum Enigma.*[4] Let me therefore commence my response by commenting on the logical structure of my argument as presented in that monograph. I begin namely, in chapter 1, by documenting the claim that *the contemporary understanding of physics is in truth based squarely on the philosophy of René Descartes,* and thus on Dr. MacAndrew's "one apple" ontology. That ontology assumes, namely, that the "real world"—measuring instruments and all—reduces to the *physical,* and thereby postulates a world

---

3. The one exception appears to be what is currently known as "superdeterminism," which strikes me however—not in fact as a solution—but rather as a *reductio ad absurdum* showing that the measurement problem admits of no solution on the physical plane.
4. Angelico Press, 2005.

which can in principle be described "without residue" in terms of physics alone. And it is that "description," of course, particle physicists are currently laboring to perfect: their objective is to specify the complete inventory of "quantum particles" out of which, supposedly, everything in the universe is composed.[5]

What I wish to point out at this juncture is that the aforesaid chapter 1 clearly explains the difference between my *Weltanschauung* and that of the physics establishment, which Dr. MacAndrew appears to make his own: in posing the question "When Is an Apple Not an Apple?" he wishes evidently to discredit my "two-Apple" ontology. I understand that. What I can't understand is how, in the course of his *contra* "two-Apple" argument, he can say *"it transpires that Smith's entire critique of 'the Cartesian assumption' is uncalled for and misdirected"*—when in fact my argument *hinges upon the denial of that very "Cartesian assumption"*! I can understand well enough that someone might disagree with my critique of the Cartesian assumption—but this is by no means what Dr. MacAndrew claims. Here is the passage:

> Smith never properly explains how accepting the idea of mind-body dualism (the Cartesian assumption as he calls it), either consciously or unconsciously, leads to all the errors of modern thinking that he claims to exorcise, and how abandoning it allows the scales to fall off from our eyes and all that is paradoxical in quantum mechanics to become clear. Smith's inability or unwillingness to set out his case clearly against "the Cartesian assumption" is frustrating. We see that he is railing against something, but he identified neither the thing nor the rationale for his opposition, nor what benefits would accrue from giving it up.

Let me note, parenthetically, that if I do engage in "railing," as Dr. MacAndrew avers, it is at least not *ad hominem*. What in truth I find *"uncalled for and misdirected"* is the trading of insults, which to my mind has no place whatsoever in scholarly discourse: it may be a bit

---

5. I will mention in passing that the project has not been going well, as I explain in chapter 4.

old-fashioned, but I don't condone the use of invective and slander, even in response to such. I shall *not* therefore engage in *ad hominem* slurs, nor respond in kind when Dr. MacAndrew alleges, for example, that "*he reserves his strongest vitriol for Descartes.*" My criticism is directed—and I should hope, without any "railing" or "vitriol" at all—not against René Descartes the man, but against his bifurcationist ontology rather, which in my view is not only misconceived, but has in fact befuddled Western civilization since the onset of the Enlightenment, to the detriment of our very humanity.

What I wish to respond to is the contention that I don't properly explain or justify my opposition to that bifurcationist ontology—what Dr. MacAndrew refers to as "the mind-body dualism" or "the Cartesian assumption"—nor what advantages accrue from its abandonment. Let me therefore answer both questions once and for all: *the reason I argue against the Cartesian postulate is that I regard it to be fallacious; and as to the benefits to be derived from this recognition, these range, I say, from the resolution of the "measurement problem" in quantum mechanics all the way to the restoration of our bona fide and untruncated humanity.* And by the way, let me note parenthetically that what MacAndrew terms the "mind-body dualism" is not in fact the same thing as "the Cartesian assumption": it suffices to note that the Thomistic ontology, for instance, affirms a "mind-body" dualism by distinguishing ontologically between *corpus* and *anima*, yet categorically rejects the so-called "Cartesian assumption." The great schools of premodern philosophy[6] concur in fact on both these points: the rejection namely of "Cartesian bifurcation" and the affirmation of a "mind-body" dualism. My "offense" seems to be that I stand with what some have termed *sophia perennis*, in diametric opposition to the current

6. I have not forgotten the case of Democritus, "the father of atomism," whose celebrated fragment expresses in a nutshell the gist of the Cartesian bifurcationist philosophy. The point is that one may conclude on cogent grounds that Democritean atomism constitutes a heterodox doctrine, rejected by the dominant philosophical schools of antiquity. What Descartes has done is to revive an ancient heresy; and whereas the leading philosophers of antiquity were wise enough to reject that doctrine, the gurus of the Enlightenment were not.

wisdom: the *Zeitgeist* of the present age which, as Carl Jung has so aptly put it, "will not let itself be trifled with."[7]

~

Out of the literally scores of objections Dr. MacAndrew has raised in opposition to my views, there is one which does in fact strike the mark: it relates to the aforesaid chapter 1 of *The Quantum Enigma*, in which I contend against Cartesian bifurcation by arguing for a realist view of visual perception. I did so as best I could at the time, and admittedly fell short of achieving the stated objective of "Rediscovering the Corporeal World." Here then is what Dr. MacAndrew has to say in that regard:

> Having examined the "ghost in the machine" or the "Cartesian the-ater" doctrine of visual perception, the idea that there is a separated mind which observes an image in the brain produced by the visual neuro-system, and having rightly found that concept wanting, he concludes "the missing piece of the puzzle must be strange. Call it mind or spirit or what you will" thereby coming full circle and plumping for a solution that is indistinguishable from the mind-body substance dualism of a distinctly Cartesian kind.

To be sure, I was by no means "plumping for" anything even remotely affiliated with a "mind-body substance dualism of a dis-tinctly Cartesian kind": I was on the contrary in quest of a *realist understanding of visual perception*, which locates the redness of a rose, neither in neurons nor in a "separated mind," but precisely *in the rose itself*, where according to common belief—in which we all share at least most of the time—it actually belongs. And this is the point regarding which Dr. MacAndrew has in fact hit the mark: I *did* in truth fail to resolve the enigma of visual perception! I failed

---

7. *Modern Man in Search of a Soul* (Harcourt Brace, 1933), p. 175. It should perhaps be noted that whereas I heartily agree with Carl Jung on this par-ticular point, I am by no means an aficionado of his psychological theories, a subject on which I have expressed myself at considerable length in *Cosmos and Transcendence* (Angelico Press, 2008), chapter 6.

thus to deliver the very thing the chapter title itself proclaims: the "rediscovery," namely, of what I henceforth termed the *corporeal* world. Let me then respond to Dr. MacAndrew's point by shedding light on this issue.

To begin with, I might note that based upon the sapiential traditions of mankind—from the Vedic to the Platonist and the Thomistic—I never seriously doubted that we do "look out upon the world":[8] the problem thus was simply to comprehend, as best I could, how that prodigy of "perceiving the external world" is actually accomplished. I realized, moreover, that there can in principle be no empirical verification of the bifurcationist postulate: that the Cartesian premise itself renders this inconceivable. It may be recalled that Descartes himself was obliged to ruminate for days, in the stillness of his garden retreat, to convince himself that an "external" world so much as exists, and was obliged in the end to invoke "the veracity of God" to persuade himself of that fact![9] In my own reflections I had the advantage of "hindsight": I could see what four centuries of "Cartesian bifurcation" have done to our civilization and our humanity, and had in fact written an entire book on that subject.[10] I realized that the Cartesian *Weltanschauung* had *de facto* given rise to a new breed of humans, what Philip Rieff terms "psychological man" as distinguished from *homo religiosus*, and characterizes as "born to be pleased."[11] Above all, however, I was loath to accept an unprovable dogma which no one—not even those "born to be pleased"—can fully and consistently believe. So I wrote my first chapter in *The Quantum Enigma*—the one entitled "Rediscovering the Corporeal World"—as best I could; and admittedly Dr. MacAndrew is right: *I did not* in fact *succeed in*

---

8. To those who might invoke *advaita* Vedanta, say, to brand this realism "naïve," let me note that what may indeed be "naïve" is not the assertion that *we do perceive this world*, but the assumption, rather, that this "perceptible world" itself constitutes the ultimate reality. The rejection of Cartesian bifurcation proves thus to be a step at least in the right direction.

9. I am reminded of the "theistic" evolutionists, who likewise invoke "God" to rescue their misconceived theory when it runs aground!

10. *Cosmos and Transcendence*, op. cit.

11. *The Triumph of the Therapeutic* (Harper, 1968), p. 13.

proving my case, be it on scientific or philosophical ground.

Little did I know, however, that someone else had in fact succeeded in this very enterprise, and *on scientific ground* no less! I am referring to James J. Gibson, a cognitive psychologist who began his career in the 1940's at Cornell University on a government grant, missioned to discover how one can visually perceive a so-called "aiming point": e.g., the deck of an aircraft carrier as viewed from an approaching plane. And this inquiry led Gibson to the fateful discovery that, based on the reigning "retinal image" paradigm for visual perception, such an aiming point *cannot in fact be perceived at all*: the information contained in "retinal images" simply doesn't suffice! At that juncture, this intrepid empiricist did the un-Kuhnian thing: he abandoned the "reigning paradigm." And thus began one of those rare and amazing "voyages of discovery" which would doubtless have gained worldwide renown if it were not for the fact that it offends against that post-Enlightenment *Weltanschauung* "which will not let itself be trifled with": the hallowed dogma of "Cartesian bifurcation" no less. Following decades of painstaking—and, I would say, *brilliant*—experimentation, Gibson was able at last to identify what it is that enables visual perception of the external world, and formulate on this basis a new paradigm which does apparently square with the empirical facts. And needless to say, in a perfect world the new paradigm would have replaced the now discredited "visual image" approach.[12]

What concerns us is the fact that the Gibsonian theory of visual perception actually affirms "the unspeakable": namely that *we do in truth perceive the "external world"*—as just about everyone, from simple folk to the great philosophers, had thought all along. And that is why Gibson refers to it as "the *ecological* theory of visual perception," the point being that what we actually perceive is not "inside the head" after all, but pertains in fact to what he terms "the environment," which proves thus to be inherently what I term the *corporeal*

---

12. No one, to be sure, denies that the "visual image" paradigm does have its domain of application: e.g., that it works just fine for the prescription of spectacles.

world.[13] The fact is that Gibson succeeded where I had failed, and did so moreover—not on philosophical—but on *empirical* ground.

⟋

I have not forgotten Dr. MacAndrew's litany of accusations, nor have I digressed from the objective of attesting that I may not be quite as misinformed, scatterbrained and dull-witted—nor even as "vitriol-spewing"—as my detractor makes me out to be. The reason James Gibson enters the picture is that he has supplied the very *"rediscovery of the corporeal world"* I had failed to deliver, and in so doing substantiates my basic claim.

It is to be noted that whereas Gibson arrived at this discovery on empirical grounds, it turns out—as I have shown elsewhere[14]— that his finding accords with sapiential tradition: for not only is the resultant theory of visual perception non-bifurcationist, but proves in fact to be inherently Aristotelian. The point is that the "ecological theory" of visual perception is based, not on "images"—be they "retinal," "neural," or "mental" (whatever that might mean)—but on what Gibson terms "invariants in the ambient optic array." And as that which connects the perceptible to the perceived—as something, therefore, which both the external object and its percipient can possess—these invariants are evidently tantamount to "forms." It appears that, in the final count, *there may be* no solution to the problem of visual perception which is not intrinsically hylomorphic.

Let this suffice to show that the "corporeal apple" may prove to be real after all, and that Dr. MacAndrew may have been a bit hasty in concluding that *"Smith's entire critique of 'the Cartesian assumption' is uncalled for and misdirected, leaving the reader with a powerful sense that the first two chapters of the book have gone awry."* The fact is that my interpretation of physics is based squarely on the denial of that "Cartesian assumption," and that this unfounded

---

13. I have dealt with the Gibsonian discovery in several publications, beginning with "The Enigma of Visual Perception," *Science & Myth: With a Response to Stephen Hawking's* The Grand Design (Angelico Press, 2012).

14. Namely in "The Enigma of Visual Perception," referenced in footnote 13. We return to this issue in chapter 5.

hypothesis, which has all long befuddled the physics community, turns out to be a fraud.

The problem, however, is that the Cartesian *Ansatz* is so deeply ingrained in the contemporary scientific mind as to be *de facto* untouchable. I am reminded of an incident sparked by a lecture I delivered at the University of Notre Dame, exposing the bifurcationist muddle in the philosophy of physics. Some weeks later I received a letter from a visiting professor I had met at the conference, informing me that he had attempted repeatedly to explain that "Cartesian hypothesis" to senior members of the Physics Department, and had finally given up: "I have become persuaded," he writes, "that those who have worked in physics for a length of time are no longer capable of grasping the idea of bifurcation."[15] I seem to be having exactly the same experience with Dr. MacAndrew.

To demonstrate as clearly as possible that the critique of what he terms "the Cartesian assumption" is *not* in fact "uncalled for and misdirected," let me explain what it is I claim to establish in *The Quantum Enigma*. Having done so rather succinctly in *Physics and Vertical Causation*, I will quote the paragraph:

> Given then that we do perceive the external world—that the grass is actually green and the red apple[16] in my hand is not after all a *res cogitans*—given this fundamental premise, I posed the question whether it is possible to interpret physics *per se*, based on its inherent *modus operandi*, in non-bifurcationist terms. And let it be noted at once that *the non-bifurcationist interpretation of physics cannot be rejected on scientific grounds*. Or to put it another way: if indeed it is possible to transact the business of physics on a non-bifurcationist basis, then—contrary to the prevailing belief—Cartesian bifurcation is bereft of scientific support. *The contemporary Weltanschauung—which implicitly assumes bifurcation to be a scientific fact—has then been disproved.*[17]

15. I am referring to Prof. William A. Wallace, O.P.
16. This could be the source of Dr. MacAndrew's "Apple Not an Apple" metaphor.
17. Angelico Press, 2019; pp. 15-6.

What concerns us is the implication that *the non-bifurcationist interpretation of physics cannot be rejected on scientific grounds*—and no amount of denigration or calumny can alter this fact.

~

I wish finally to return to the measurement problem in quantum mechanics to show that whereas it has proved in effect to be unsolvable on a bifurcationist basis, the new *Ansatz*—i.e., that we perceive the *corporeal* as distinguished from the *physical*—leads directly to its solution. For inasmuch as the act of measurement must terminate in a *perceptible* state of the measuring instrument, the aforesaid recognition entails that *the measuring instrument must be corporeal.* It follows that *the denial of the corporeal renders the measurement problem insoluble*: the matter is basically as simple as that!

The process of measuring a quantum system involves thus an interaction between a *physical* system and a *corporeal* measuring instrument, which cannot be explained in quantum mechanical terms: hence the so-called "collapse" of the wave function. It should hardly come as a surprise, moreover, that the quantum system itself—with its multilocating so-called "particles"—is not perceptible. As has often enough been noted, what exists prior to measurement are not actual particles, but *potentiae* represented by a probability distribution, actualizable through an act of measurement. As Heisenberg noted at the very outset, the entities described in quantum theory, such as atoms or elementary particles, "form a world of potentialities or possibilities rather than one of things and facts."[18] The moment, therefore, that one has arrived at a realist view of visual perception, the matter is settled: the measuring instrument does not—cannot then—reduce to "atoms or elementary particles." The act of measurement entails therefore a transition from potency to act, and thus from one ontological plane to another.

However, a transition between two ontological planes must be instantaneous, and I will note that this fact has major implications: for it happens that the forces known to physics do not account for

18. *Physics and Philosophy* (Harper & Row, 1958), p. 186.

such an instantaneous effect. It turns out thus that the act of measurement brings into play a hitherto unrecognized mode of causality, which acts—not by way of a transmission of some kind—but instantaneously: what I term *vertical* as distinguished from *horizontal* causation, the kind that acts by way of a temporal process. It happens, moreover, that the newly discerned causation accounts for various phenomena which have baffled physicists, beginning with *nonlocality*: what Albert Einstein referred to skeptically as "spooky action at a distance." However, what renders vertical causation "spooky" is simply the fact that it does not act "in time" and consequently does not "propagate" through space. Meanwhile, however, we have come to know, on the strength of William Dembski's now famous 1998 theorem—subsequently associated with the epithet "intelligent design"—that it takes in fact vertical causation to produce "complex specified information": which means, for example, that in writing this book I have availed myself of VC.

Getting back to the measurement problem, from which I have digressed: the fact is that the Schrödinger equation breaks down—is "re-initialized" as the physicists say—at the very instant of measurement; and the question is, "Why?" What is it about the measuring instrument that enables it to interrupt the Schrödinger evolution of the system? The point is that if the measuring instrument were itself a physical system—if the "one-apple" hypothesis were true—that discontinuity could not occur. Not only, therefore, does the ontological distinction between a corporeal object and a quantum system resolve the measurement problem, but it happens that *nothing else will*: there *is* no other way out of the quandary. Instead of my critique of the so-called "Cartesian assumption" being "*uncalled for and unnecessary*" as Dr. MacAndrew charges, it turns out that *nothing less than the categorical denial of that very assumption will resolve the quantum enigma.*

~

Enough has been said, I trust, to refute the central objection brought against me by Dr. MacAndrew. I deem it to be unnecessary, therefore—not to say a waste of everyone's time—to respond to the thirty or so remaining accusations by enlarging this "brief" to a "full" response.

2

THE TRIPARTITE
WHOLENESS

THE UNIVERSE PRESENTS itself as extended in space and changing in time, and its spatio-temporal locus can be represented mathematically in terms of three spatial coordinates and one temporal: the familiar quadruple (x_1, x_2, x_3, t). The "container" is thereby decomposed conceptually into an infinite aggregate of punctual loci, which define a corresponding decomposition of the "content": and this twofold conceptual atomization constitutes the foundation upon which physics as such is based. The fact that it operates by way of differential equations is thus predetermined, as is presumably a significant part of what we normally take to be its "empirical findings." What needs to be grasped is that physics deals not simply with the universe as such—as almost everyone assumes—but is descriptive rather of the cosmos as "atomized" by way of the aforesaid spatio-temporal fragmentation. It is hardly surprising, therefore, that the manifestations of authentic wholeness—from the simplest to the most profound—prove as a rule to elude its grasp.

Metaphysics, on the other hand, would know that which precedes this fragmentation ontologically. It is possible of course to suppose that there is no wholeness at all—that the "atomized" universe is all there is—and ontologies of that kind have been enunciated, time and again, beginning perhaps with Heraclitus. Yet the fact remains that the great sapiential traditions, both of the East and the West, have emphatically declared the contrary: i.e., that *wholeness precedes division*—not temporally, to be sure, but *in veritas.* What *is* transcends both spatial and temporal division: and it is upon this very recognition, precisely, that authentic ontology rests.[1]

Given thus that the primary cosmic reality is both supra-temporal and supra-spatial, one might imagine that the spatial and temporal bounds definitive of the cosmos, as it normally presents itself to mankind, are imposed upon that antecedent wholeness "at a single stroke" as it were. There are however cogent grounds to conclude that such is not the case, and that in fact *the bound of time precedes the bound of space*—not temporally, to be sure—but in a metaphysical sense: the integral cosmos proves therefore to be *ontologically tripartite.* Between the pre-temporal wholeness and the spatio-tempral world, namely, there exists a domain subject to the bound of time but not of space; and let me note that this metaphysical fact has scientific implications, the most evident being that it disqualifies Einsteinian physics at a single stroke. For in distinguishing categorically between "time" and "space," the given trichotomy affirms the existence of a universal temporal "now": an absolute simultaneity namely, defined throughout the length and breadth of the corporeal universe—which the Einsteinian construct of "space-time" rules out.

I would note[2] that the concept of a tripartite cosmos is indigenous to the sapiential traditions of mankind, and has apparently received its most explicit formulation in the Vedic, where the cosmos thus conceived is known as the *tribhuvāna* or "triple world." Inasmuch, moreover, as the discernment of the *anthropos* as a tripartite microcosm—composed of *corpus, anima,* and *spiritus*—was current in Europe right up to the Enlightenment, the conception

---

1. For those who respect the Judeo-Christian tradition I would point out that the *"ego sum qui sum"* of Exodus 3:14 suffices to make this abundantly clear. Moses asked by what "name" God is to be known, and the answer he received was: *"I am who am."* We are not suggesting, to be sure, that the cosmos shares the eternity of God, but only that, on its highest plane, it likewise transcends the flux of time. This is what in Scholastic parlance was referred to as *aeviternity.*

2. See *Physics and Vertical Causation: The End of Quantum Reality* (Angelico Press, 2019), pp. i–vi.

of a *tribhuvāna* is in a way indigenous to the West as well. This isomorphism between man and cosmos taken in their integrality—which likewise pertains to the sapiential traditions—entitles us to designate the primary cosmic domain as the "spiritual," and the third as the "corporeal." The second could of course, by the same token, be termed the "animate," yet for reasons which need not detain us, we shall refer to it as the "intermediary" domain.[3] But whereas the concept of the *tribhuvāna* proves to be well documented in the traditional literature, its specification in terms of spatial and temporal bounds, as given above, seems to be extant neither in ancient nor in contemporary sources. What primarily concerns us, however, is not the historical origin, but the scientific and ontological implications of the given trichotomy.

The most basic and indeed most obvious conclusion to be drawn is that physics as such is restricted to the lowest of the three cosmic domains: to the *corporeal* namely—for the simple reason that the spatial bound ceases to apply above corporeal domain. The intermediary and the spiritual are consequently "invisible" to the physicist, restricted as he is in his purview to regions within space and time.

Having thus recognized its proper domain, we need to take note of the fact that physics entails further restrictions, that in the final count it "has eyes" only for spatio-temporal bounds: for the "container" namely, as distinguished from its "content." And what *is* that "unperceived" content? It follows from our conception of the tripartite cosmos that this content pertains ultimately to the

3. This is what, in nineteenth-century occultism, was known as the "astral plane," and what the Orthodox Church refers to as the "aerial world." But whereas nowadays this realm is virtually unrecognized and unaccessed in the West, it can in fact be entered and crossed by means which may in principle be termed "scientific": the catch is that when it comes to sciences of this kind—which have existed in the past, and may still be operative in some remote corner of the globe—the "instrument" can ultimately be none other than the man himself.

spiritual realm: the primary reality from which both the intermediary and the corporeal domains derive. The true "substance" of all entities—be they corporeal or intermediary—pertains thus to the spiritual order. It can also be said, however, that the "substance" of a corporeal entity pertains to the intermediary domain: the "subtle"[4] world which transcends the spatial but not the temporal bound: for inasmuch as the "substance" of a *subtle* entity pertains evidently to the spiritual domain, there is in fact no contradiction between these respective conceptions.

It is to be noted that the "subtle substance" of a corporeal entity corresponds to the Thomistic conception of a *substantial form*. I say "corresponds," because the point of view implicit in our conception of "tripartite wholeness" does not coincide with the hylomorphic perspective: what we conceive as the "unbounded," after all—so far from reducing to the nonentity of *materia prima*—pertains in fact to the highest level of cosmic reality. Yet it turns out that the tripartite and the Thomistic ontologies are closely related, to the point that the "subtle substance" of a corporeal entity in the tripartite ontology is tantamount to a substantial form, as we have said.

∿

Physics deals ultimately, as we have noted, with functions $f(x_1, x_2, x_3, t)$ which bear reference to *corporeal* reality, howbeit by way of two stupendous reductions: first, a decomposition of the "container" into an infinite set of infinitesimal fragments, each defined by four coordinates (x_1, x_2, x_3, t), followed by a quantification of the thus-fragmented content, which reduces that "infinitesimalized" constituent in effect to a *density* of some kind, defined in appropriate units according to the cgs or "centimeter-gram-second" system of standard magnitudes.

At first glance one cannot but be amazed that a strategy so disintegrating—and so "superficial"—can give rise to a science of immense efficacy, let alone to the discovery of so-called "laws of

4. As equivalent to the Sanskrit "*sūkshma*" by which the intermediary region of the *tribhuvāna* is characterized in the Vedic tradition.

nature": *the* laws of nature no less, as we are told. Admittedly there are factors in play which favor the enterprise, the most obvious perhaps being that, under laboratory conditions, the system is prepared to conform to the very criteria of "uniformity" upon which physics as such is based. So too, in applications of physics to natural objects, the effects of nonuniformity may cancel on statistical grounds, as happens for example in the Newtonian calculation of planetary orbits, in which an entire planet, with all that it carries, is conceived as a single "mass point" located at its center. Yet it needs to be understood that considerations of this kind hardly suffice to account for the possibility—let alone the staggering "multi-decimal" accuracy—of mathematical physics.

This brings us to the question of causality, and the point I wish to make above all is that there exist perforce two kinds: what I term the "horizontal" kind, which operates "in time" by way of a spatio-temporal process, and the "vertical" kind, which does not. But whereas it is generally assumed that horizontal causation covers the entire ground, the opposite has now come to light: for given that the primal substance is subject neither to space nor to time, *the absolute primacy of vertical causation* is obvious from the start.

It is apparent, first of all, that by virtue of acting "above time,"[5] vertical causation originates in the highest ontological sphere: the one that is not subject to the bound of time. Yet inasmuch as vertical causation may act directly or by way of a substantial form, two possibilities present themselves on the lower planes. As regards the direct mode of action, this corresponds evidently to what St. Thomas Aquinas terms "the act of being," conceived as "*the most intimate element in anything, and the most profound element in all things, because it is like a form in regard to all that is in a thing.*"[6] Yet it must not be supposed that individual beings have no efficacy of their own: the fact is that every corporeal entity, in particular, endowed as it is with a substantial form, does in truth have the capacity to act by way of vertical causation, a power which

---

5. Regarding the instantaneity of vertical causation, I refer to *Physics and Vertical Causation*, op. cit., pp. 26-9.

6. *Summa Theologiae* I, Q. 8, Art. 1.

evidently attains its zenith on the human plane in what are traditionally termed acts of "free will." And let me add parenthetically that the existence of such "vertical" acts can in fact be established rigorously on the strength of William Dembski's so-called "design inference" theorem,[7] which affirms, on mathematical grounds, that the production of what is termed "complex specified information" or CSI cannot be effected by way of horizontal causation alone.

~

Given that the primary mode of causality is vertical, it follows that horizontal causation cannot but be an effect of that vertical causality. There is thus a hierarchic distinction between the two: the vertical holds primacy. So too their respective modes of action are vastly different, and in a way complementary: the vertical acts instantaneously, as we have said, and over unlimited regions of space, reflective of its supremacy, whereas the horizontal operates "laboriously" one might say, by way of a temporal progression through space. The fact is that vertical causality derives from wholeness, whereas the horizontal derives from parts: and ontologically speaking, *the whole has primacy over the parts.* And this ontological recognition has scientific implications: it entails namely that *horizontal causality applies to the extent that it is not "over-ruled" by the vertical.* It applies thus in the limiting case in which effects of wholeness can be neglected. To which I would add that this way of looking at the matter has in fact been pioneered by the physicist David Bohm, who arrived at the realization—so very startling to the Einsteinians—that the breakdown of horizontal causation in the phenomenon known as *nonlocality* constitutes in truth an effect of what Bohm himself refers to as "wholeness."[8]

To comprehend a science, it is above all needful to perceive its limitations; and it should be clear from the start that in the case of physics, the prime limitation stems from the fact that, by its very

---

7. See *Ancient Wisdom and Modern Misconceptions* (Angelico Press, 2013), pp. 180-93.
8. See ch. 6.

modus operandi, it obliterates *wholeness* in the most drastic manner conceivable. The wholeness of a corporeal entity, however, derives from its substantial form, which makes it incumbent to recognize that there are in fact different *kinds* of substantial forms. The most basic—and doubtless the most useful—classification is given by the time-honored distinction between *mineral* or inorganic, *vegetative*, *sentient*, and *rational* substances: we may think of these designations as defining four different degrees or levels of wholeness, and thus by implication, four corresponding "grades" of vertical causation.

A number of corollaries present themselves almost immediately, beginning with the realization that if horizontal causality fails to cover the entire ground in the inorganic domain—as the phenomenon of nonlocality demonstrates—what to speak of even the lowest levels of the organic, in which wholeness plays an incomparably greater role! It is hardly surprising, therefore, that by the time we arrive at the wholeness of *homo sapiens*, the preponderance of vertical causation stares us in the face, beginning with the prodigious output of CSI characteristic of the *anthropos*.

This leads quite naturally to the conclusion that horizontal causality applies on the corporeal plane in the limit when the effects of wholeness are small enough to be negligible. Inasmuch, however, as it pertains to the very nature of the organic "whole" to control the structure and functioning of every part, down to a single cell, it is evident that the aforesaid condition is nowhere met in the organic realm. It follows that physics can in truth be no more than *the science of the inanimate*—to which one should add that even in the inorganic realm, as we have noted, effects of "wholeness" impede the hegemony of horizontal or "mechanistic" causality.

∿

Pre-quantum physics was said to be the science of "matter," a hybrid notion which proves to be murky in the extreme. What concerns us are two key points: first, that one arrives at this idea of "matter" by casting out all that traditional ontology knows as "form" or "substance"; and secondly, that this so-called "matter" is conceived to be inherently "granular," which is to say that it consists supposedly of minute and indecomposable particles termed *atoms*. To be sure, there is a certain logic in these suppositions inasmuch as they conform evidently to the *modus operandi* of physics. Yet from the standpoint of traditional ontology, the conclusion imposes itself that such a proposed reduction of the whole to its "atomic" parts is stringently wrong-headed: that ontologically speaking, the physicist is actually looking "downwards" in the direction of "outer darkness," where all *being*—and consequently all *knowing*—come finally to an end.

It is safe to say that the discovery of the quantum realm in the early decades of the twentieth century came as a complete surprise to both sides: to the physicist, because what he discovered proved to be "sub-existential," and to the philosopher, because he never expected that a *materia quantitate signata*[9] could attain even that much "empirical" reality, let alone give rise to a quasi-magical technology: to *"signs and wonders that could deceive even the elect"*! Yet even so the philosopher's conclusion stands, now as before: the physicist *is* looking "downwards," in the direction of *materia prima* where nothing—not even a "quantum particle" of any kind—exists. And though it turns out that "between" the corporeal world and that "nothingness" there are atoms composed of protons, neutrons, and electrons—quantum particles which "half-exist" if you will—it has since become clear that continuing "downwards" one arrives at "depths" where not even phantom particles such as "quarks" are to be found: the recent history of particle physics should suffice to make this abundantly clear.[10]

Yet one sees at the same time that the physicists have made a stupendous discovery: midway, namely, between the corporeal world and "outer darkness"—which cannot after all be exorcised—they have detected entities which seem to manifest a kind of weird reality of their own. The catch, however, is that this "world" does

---

9. An expression used by St. Thomas Aquinas apparently to designate a "sub-existential" entity defined in purely quantitative terms.
10. See ch. 4.

not in truth exist![11] One finds in the end that the quantum realm consists—not of actual entities—but of *potentiae* that can be actualized on the corporeal plane in basically three ways: by way of *measurement* namely, along with what I have elsewhere termed *presentation* and *display*.[12]

It is to be noted that a seeming discrepancy with our "tripartite" conception of corporeal being has thus come into view, which we need forthwith to clarify: for whereas we have conceived of the corporeal domain as simply the cosmos subject to the bounds of space and time, we are now confronted—within this very space and time—by a "subcorporeal" realm, which in a sense "underlies" the corporeal. We need thus to sharpen our conception of the latter: it no longer suffices to define corporeal being by its subjection to the bounds of time and space. What is it, then, that differentiates a corporeal entity, properly so called, from a conglomerate of so-called quantum particles? And by now the answer to this question is not far to seek: it is—and cannot but be—the fact that *corporeal entities own a substantial form*—which is tantamount to saying that *they exist*.

~

Long ago Richard Feynman observed that "no one understands quantum theory"—and this statement proves to be profound: how, after all, *can* one understand a "world" that only half-exists! There are those, to be sure, for whom it suffices that quantum theory "works"; yet, even so, "no one understands."

What readily can—and *should*—be understood, on the other hand, is that the quantum world does not exist "by itself," in

---

11. I have dealt with this issue extensively, beginning with *The Quantum Enigma* (Angelico Press, 2005). For a brief introduction I refer to my article, "From Schrödinger's Cat to Thomistic Ontology," *Ancient Wisdom and Modern Misconceptions*, op. cit., chapter 1.

12. These are technical terms introduced in *The Quantum Enigma*, the first of which refers to the corporeal object X represented by a corresponding quantum mechanical system SX, and the second refers to the visible representation of a physical process or entity, as in an oscilloscope or a cloud chamber.

separation from the corporeal: to think that it does is to ascribe to the quantum realm a reality which in truth it does not possess. And once again it is Werner Heisenberg who has given us the requisite clue when he observed that "science [meaning physics] no longer stands before Nature as an onlooker, but recognizes itself as part of this interplay between Man and Nature."[13] One might add that the same conclusion, inherently, had previously been reached by Sir Arthur Eddington, who went so far as to surmise that "the mathematics isn't there until we put it there"—to which, in a poetic vein, he adds: "We have discovered a footprint in the sand; and lo, it is our own!"[14] Regardless, however, of whether "the mathematics" is or is not there until "we put it there," my point is that Eddington, no less than Heisenberg, perceives physics as an "interplay between Man and Nature."

That "interplay," however, cannot be effected without vertical causation; and the easiest way to recognize this fact is to reflect upon the act of measurement which connects the physical to the corporeal. We have here a transition from the physical to the corporeal plane, and it is self-evident that a transition between two distinct ontological planes cannot but be "instantaneous." One finds that horizontal causality is simply not up to the task. Quantum measurement proves thus to be an incurably vertical act—which is of course precisely the reason why the so-called measurement problem turns out to be insoluble on the plane of physics. The fact is that the "collapse" of the wave function cannot be understood in terms of the Schrödinger equation, to put it in quantum-mechanical terms: for an instant, literally, that differential equation is superseded, and this instantaneity constitutes the "hallmark" of the vertical. As the Scholastics might say, vertical causality acts in the nunc stans: in the "now" that stands.

~

13. Das Naturbild der heutigen Physik (Hamburg: Rowohlt, 1955), p. 21.
14. The Philosophy of Physical Science (Cambridge University Press, 1939), p. 137.

Having once recognized the necessity of vertical causality in the economy of physics, it becomes apparent that such acts of causation come into play in a wide range of phenomena known as *quantum entanglement*, in which groups of quantum particles are mysteriously connected, so that the results of measurement, applied to individual particles within the group, are "instantaneously" correlated. Entanglement results thus from a certain *wholeness* which cannot be understood in terms of horizontal causality, and must consequently be ascribed to a vertical mode of causation—what Einstein famously ridiculed as *spukhafte Fernwirkung*: "spooky action at a distance."

Entanglement ceases of course to be "spooky" above the corporeal plane: for where there is no more distance, there obviously is no "action at a distance" as well. The point is that vertical causation—acting, as it does, from the primary level of the cosmic trichotomy—"sees" no distance, if one may put it so, and consequently has no distance to cross. Entanglement then is caused by an interaction between the quantum system and the corporeal world in that "interplay between Man and Nature" from which physics as such derives, which involves acts of vertical causality.

What appears however to have puzzled physicists more even than quantum entanglement is its cessation on the corporeal plane. Once one assumes that the cosmos at large reduces to an ensemble of quantum particles, "decoherence" turns into an insoluble mystery: in a universe in which particles multilocate and cats can be both dead and alive, the world as we know it is thus in effect inexplicable. The customary answer, to be sure—the one that comes instantly to mind—is that the bizarre features of the quantum realm, such as multilocation, cancel out statistically when it comes to ensembles large enough to be actually perceivable—which proves however not to be the case: this is, after all, what Schrödinger proved by way of his famous cat. The truth of the matter is that so long as all things reduce to quantum particles, the weirdness of the quantum realm bleeds inevitably into the macrocosmic domain.

It is imperative therefore to realize that "all things" do *not* in fact reduce to quantum particles. Yet how many physicists will believe you if you were to suggest that there may indeed be "*more things in heaven and earth, Horatio, than are dreamt of in your philosophy*"! What needs to be grasped is that the corporeal realm is literally "worlds removed" from the inherently mathematical conceptions to which physics has been reduced since the advent of quantum theory. As Sir Arthur Eddington may have been the first to observe, it is at that juncture—that point of transition—that "the concept of substance" disappeared from physics. The crux of the matter is that this "disappearance of substance" is what defines the quantum realm and entails that the latter does not in truth "exist": that it consists—not of actual particles—but of *potentiae* "midway between being and nonbeing" to quote Heisenberg again.

The attempt to understand how the corporeal world arises from the physical—what is nowadays conceived as "decoherence"—cannot but falter, for the simple reason that the corporeal world does *not* in fact "emerge" from the physical. It is actually the other way round: for as Heisenberg and Eddington have recognized, it is "the interplay between Man and Nature"—*corporeal* Nature, in particular—that gives rise to the quantum realm. Attempts to explain the corporeal domain in terms of physics—and hence in terms of horizontal causation—prove thus to be wrong-headed; and no one among contemporary physicists seems to have grasped this more sharply than Steven Weinberg. "You can very well understand quantum mechanics," he writes,

> in terms of an interaction of the system you are studying with an external environment which includes an observer. But this involves a quantum mechanical system interacting with a macroscopic system that produces the decoherence between different branches of the initial wave function. And where does that come from? And, strictly speaking, *within quantum mechanics itself there is no decoherence.*[15]

15. Quoted by Sabine Hossenfelder, *Lost in Math* (Basic Books, 2018), p. 126; italics mine.

And the reason, moreover, why "there is no decoherence" in the quantum realm has by now become evident: it resides in the fact that decoherence demands *vertical* causation, which normally acts by way of a substantial form. There is no decoherence therefore "within quantum mechanics itself," for the simple reason that, in the quantum realm, *there is no* substantial form.

$\sim$

Finally let us consider whether indeed "the mathematics isn't there until we put it there" as Eddington claims. It is to be noted, first of all, that the quantum realm as such is an effect—not of horizontal—but perforce of vertical causality, for as we have noted in reference to the measurement problem, horizontal causality is restricted to a given ontological plane. Inasmuch as that vertical causality, moreover, arises from "an interplay between Man and Nature," it arises not simply from Nature alone. The point is that Man—"we" as Eddington has it—have a part to play, and quite clearly it is the active part: it is "we" who ask the questions so to speak, and it is Nature that gives the response. And that "response" *is* in effect the physical.

How, then, do we "ask the question"? We do so, clearly, in the capacity of the experimental physicist: by means of corporeal instrumentation that is. Yet the essential ingredient proves to derive from the theoretician: it is he who—by means of the experimentalist—"poses the question" that elicits the answer, and in so doing "puts the mathematics there" as Eddington has it. Yet even so it is Heisenberg who has the last word: for it is not actually *we* who "put the mathematics there," but "Nature" that does so—howbeit in response to *our* "question."

This leaves open, of course, whether Eddington's epistemological and group-theoretic derivations of that "mathematics" prove to be cogent. The fact that he calculates the fine structure constant, for example[16]—without reference to a single measurement—to be

16. *Relativity Theory of Protons and Electrons* (Cambridge University Press, 1936).

1/137, which moreover agrees with the empirical value to within three-hundredths of one percent, is of course astounding; yet, to my mind at least, reasonable doubt remains.

But be that as it may, the point is that so long as the quantum realm itself is defined through an "interplay between Man and Nature," its "mathematics" too cannot but be "put there" by means of that "interplay." We are thus confronted by the momentous realization that the "footprint" we have discovered "in the sand"—proves, in the end, to be in a way "our own."

3

FROM COSMOS
TO MULTIVERSE:
THE OMINOUS DESCENT

HAVING RECOVERED the conception of the tripartite cosmos indigenous to the sapiential traditions of mankind,[1] and noted that it relegates the corporeal world to the lowest of three ontological tiers, let us recognize that this traditional ontology entails the very opposite of what our contemporary civilization believes on purportedly scientific grounds. In the tripartite cosmos, namely, space and time cease to be the *container*, as our *Weltanschauung* deems them to be, but prove rather to be themselves the *contained*. One of the two worldviews has evidently stood the cosmos on its head: and I shall argue that it is our own. I contend that it is our fundamental science—physics namely—that has misled us, and propose now to examine its current impasse within a great arc associated, in "descending order," with three illustrious names: Plato, Aristotle, and lastly, *Galileo*.

Let us begin with the observation that the aforesaid distinction between the supra-temporal and the temporal domains accords with the dichotomy between the *intelligible* and the *sensible* worlds definitive of Platonism: ontology *per se* hinges ultimately upon that cut! The fact, moreover, that we subsequently divide the latter into two subdomains—the corporeal or "gross" and the intermediary[2] or "subtle"—is also crucial, and accords likewise with the sapiential

1. See *Physics and Vertical Causation* (Angelico Press, 2019), pp. 101-22.
2. This ontological domain is subject to the condition of time but not of space, and is consequently situated "above" the corporeal.

traditions, beginning with the Vedic, which speaks familiarly of the *tribhuvāna* or "triple world."

A glimpse, at least, of what the Platonist "primacy of the *intelligible*" entails may be gleaned from Plato's celebrated "myth of the cave": picture a group of prisoners in a cave, constrained to gaze upon shadows projected upon its wall by objects "outside the cave."[3] Certainly the objects thus perceived are, in a sense, "unreal." Yet this is not in fact the point: the decisive fact definitive of Platonist ontology is not that the sensible object is unreal or illusory, but that in truth *it has no separate existence of its own*. The "shadow," after all, has yet a connection with the "original," a nexus which ontologically supersedes its aspect of difference. One may say— following the French philosopher Jean Borella—that the sensible object constitutes a *sign* signifying an *intelligible referent*, which as such has a reality *sui generis*: i.e., a *semantic* reality, properly so called. In this optic, which I take to be authentically Platonist, a corporeal entity is not simply a "thing"—does not, in other words, reduce to its manifestation in space and time—but like the "shadows" in Plato's cave, points beyond itself to a referent transcending its spatio-temporal bounds. In the fullness of its being, therefore, that sensible object is *more* than simply "corporeal" by virtue of the fact that it transcends its spatio-temporal locus *semantically*. It has thus become apparent that *Platonist ontology entails a "vertical" dimension invisible—"by definition" one might say—to physics as such.*

Let this much suffice as an initial exegesis of Plato's "myth": that incomparable "icon" of the *sensible* vis-à-vis the *intelligible* tier of the tripartite cosmos. A "first" exegesis, I say: because, to be sure, there is always more! Authentic myth speaks in a way of *the ineffable*: for as Ananda Coomaraswamy apprises us, such so-called myth "embodies the nearest approach to absolute truth that can be stated in words."[4]

∾

3. Readers acquainted with what I term the "cosmic icon" (a circle in which the center represents the intelligible, the interior the intermediary, and the circumference the corporeal world) will recognize that the circle bereft of its center is also in fact symbolic of this "cave," the circumference representing its "wall."

4. *Hinduism and Buddhism* (Greenwood Press, 1971), p. 33.

Platonist ontology is based squarely upon the recognition that the sensible object—though but a "shadow" in itself—yet "presentifies" its intelligible prototype. And whereas the senses as such apprehend no more than its outer manifestation—the "shadows" as Plato has it—the intellect[5] has the capacity to "penetrate" the sensible image as such, and in so doing attain to the apprehension of the intelligible object itself: for it happens that *intellect* as such transcends both spatial and temporal bounds. It thus corresponds, in the human being, to the intelligible realm of the integral cosmos: we must remember that, according to the Platonist anthropology, the *anthropos* himself constitutes a tripartite microcosm, culminating in *pneuma*: *intellect*, that is.

Two comments are called for at this point. First, that this "*pneumatic*" apprehension of the intelligible world admits of countless degrees, and cannot as a rule be achieved in its higher modalities without a prolonged askesis, accessed by way of initiatic means. And the reason for this incapacity, let me add, has in fact been clearly enunciated by St. Paul when he referred to man in his present state as a *psychikos anthropos*,[6] the point being that, as such, he has forfeited the full and proper use of his highest faculty. In light of sapiential tradition, the task of authentic philosophy may thus be conceived as a restoration of man to the full use of his native faculties: and on this point, at least, the savants of Platonism and of Christianity appear to be in full accord.[7]

My second comment pertains to Jean Borella's symbolist or "semantic" interpretation of the *sensible* as a referend of its *intelligible* prototype. It appears to me that this in a way explicates what in the Platonic dialogues is conveyed implicitly by two complementary means: first by way of a dialectic, in which the Platonist

5. *Pneuma* as opposed to *psyche*, corresponding thus to *spiritus* in the traditional *corpus-anima-spiritus* trichotomy. And I would add that since the advent of the Enlightenment, the categorical distinction between *pneuma* and *psyche* appears to have been roundly forgotten in the Western world.

6. 1 Cor. 2:14

7. It is needless to point out that this "restoration" makes no sense whatever from a Darwinist point of view. But then: neither does Darwinism from a Platonist!

conclusion is implied through a kind of *reductio ad absurdum* of the opposing view; and also "allegorically," as in the "myth" of the cave. It is of course to be supposed that more than this was revealed "inside" the Academy. My point, in any case, is that Borella's explication of the sensible object as a *sign*—connected *semantically* to its intelligible referent—may help to open doors bolted shut generations ago.

~

The perceptible cosmos—the corporeal world, first of all—proves thus to be inherently *iconic*. Remember that it is the bounds of *time* and *space*—corresponding to "the wall of the cave"—which, as it were, convert the *intelligible* into the *sensible* in both its subtle and corporeal manifestations. In this optic both tiers are consequently a *semantic*—as distinguished from a *substantial*—reality: a *sensible* sign, that is, expressive of an *intelligible* prototype.

And therein lies the categorical difference between the Platonist and the Aristotelian ontologies: what in a way replaces the intelligible world in the philosophy of Aristotle—which already includes what may rightfully be termed a *physics*—is the idea of *substance*, based upon the concept of *substantial form*. It needs thus to be understood that the Platonist and the Aristotelian worldviews differ to the point of constituting two distinct "perspectives" or *darśanas*, as the Hindu philosophers would say. But whereas both viewpoints have undoubtedly their place and their validity, it is undeniable that, *qua* ontology, the Platonist takes precedence over the Aristotelian. And let us, at the same time, note that the Platonist ontology comes within a "hair's breadth" of the Christian, which perceives the visible cosmos inherently as a theophany: "*For the invisible things of Him from the creation of the world are clearly seen, being understood by the things that are made...*"[8]

8. Rom. 1:20. There is a categorical distinction, to be sure, between the Platonist and the Christian understanding of cosmic reality; yet I venture to suggest that the Platonist comes closer than any pre-Christian ontology—inclusive perhaps of the *advaita* Vedantic, which in distinguished circles is currently viewed as supreme.

The reality of the "visible" world—be it corporeal or subtle[9]—may therefore be conceived, in both the Platonist and the Christian traditions, as inherently *semantic*: visible entities, thus, are fundamentally *signs* signifying an intelligible referent. And I surmise that no philosopher in our time has understood this Platonist and Christian truth more profoundly—nor written on this subject with greater perspicacity—than Jean Borella, whose lead in this regard I propose to follow.[10] Let us begin, then, with two "Borellan" propositions, which I will quote verbatim:

> Every symbol only signifies through presentifications, that is through a participative correspondence with the realities it signifies.

> The visible world, the forms of which constitute the various symbolic signifiers, only possesses a deficient being, an image or manifestation of a world invisible and alone truly real, or at the very least, nearer to the unconditioned Source of being and truth.[11]

Here, in these two crystal-clear assertions, we have, I surmise, a master key to the Platonist ontology: the Platonist vision of cosmic reality, as distinguished, first of all, from the Aristotelian. And I would note that this Platonist ontology is precisely what the cosmic icon symbolizes, and therefore, in a sense, permits us to "see": for inasmuch as the intermediary and corporeal domains are differentiated from the primary through the imposition of bounds, the reality presentified by a *sensible* entity—which thus precedes

9. In the sense of the Vedic *sūkshma*: i.e., as pertaining to the intermediary domain.

10. An excellent place to begin the study of Borella's writings, for the English-speaking reader, might be *The Crisis of Religious Symbolism* (Angelico Press, 2016). I would point out that the book ends with a major essay entitled *Symbolism & Reality: The History of a Reflection*, in which Borella recounts the unfolding of his thought: I would recommend this as the perfect starting point. In a world in which academic philosophy has attained what might perhaps be termed the zenith of mediocrity, Borella's opus stands out as a beacon of light.

11. Op. cit., p. 78.

these bounds ontologically—coincides *ipso facto* with its *intelligible* referent, represented by the central point.[12]

~

Our ultimate objective is to examine the physics initiated by Galileo in light of Platonist ontology; and this inquiry—so far from constituting an "academic exercise"—turns out in fact to be crucial to the resolution of the impasse in which fundamental physics has now found itself for the better part of a century. The fact is that contemporary physics rests upon a concomitant ontology—unacknowledged though the fact may be—which moreover proves ultimately to be spurious. There *is* in truth no such thing as a purely operational or "positivistic" physics, if only because the physicist does not reduce to an automaton or a computer. He may indeed be schizoid—may believe, one moment, that "the grass is green," and the next that it has no color at all—but yet he is *human*, which entails that physics exists *perforce* within the ambience of a *Weltanschauung*. And whereas that worldview is reputedly based upon the "hard facts of science," it is that *Weltanschauung* itself that ultimately determines what is or is not such a "hard fact." In the final count we are all metaphysicians, whether we admit to it or not; and I find it ironic that those who pretend to despise metaphysics, prove as a rule to be justified in respect to their own.

The point is that our present-day "scientistic" metaphysics misconceives the relation between "appearance" and "reality" to the extent of actually inverting it. One fails to recognize the ontological distinction between "center" and "circumference"—between *being* and *appearance* that is—which as Borella explains, is by no means "a sundering and entails no heteromorphism": to perceive

---

12. The ontological reason, moreover, why that "point" is unique—why, on the highest level of reality, all "centers" coincide—resides again in the fact that, in the intelligible realm, there exists neither spatial nor temporal separation. As Dante has put it: "There every *where* and every *when* are focused" (*Paradiso* 28:41). One sees that the uniqueness of the central point in the cosmic icon is expressive of a profound ontological truth: the Platonist, namely, as distinguished from the non-Platonist, beginning with the Aristotelian.

it as such is to deny the semantic capacity of the cosmos, that is to say, *its very nature.* "Quite the contrary," Borella continues:

> *appearance is the image and revelation of being.* It conceals this only if we idolize it, attributing to it a reality for which it is unsuited.[13]

To which he adds, in overtly Christian terms:

> But symbolism is there precisely to awaken us to an awareness of the Other World, and to save this world by revealing its theomorphism.

History, on the other hand, reveals a human propensity—in the Occidental world especially—to rebel against that immemorial symbolism: to raze the "upward pointing" monuments, and in the name of "progress" shift the collective gaze "downwards." A man-made symbolism emerges thus to challenge and displace the archetypal; and this "anti-*mythos*" in turn reacts upon the *Zeitgeist* to exacerbate its "progressivist" tendency. Our contemporary creed bears witness to this ongoing process, yet misconstrues its significance to the point of inversion.[14] Meanwhile the onslaught of "Progress" continues to gain intensity somewhat like the winds of an approaching hurricane: and that is where we find ourselves today, for the most part without so much as a clue as to where we stand, and what has brought us there. For it happens that the very "signs" in terms of which our present bearings could be identified pertain to a culture which over several centuries has been targeted for destruction by the presiding elites.

∾

13. Op. cit., p. 78; italics mine.
14. See "'Progress' in Retrospect," *Cosmos and Transcendence* (Angelico Press, 2008), ch. 7.

What is it then, in essence, that has come to pass? The most profound answer, I believe—and certainly the shortest—is the one proposed by Jean Borella: what has taken place over the millennia of human history is *"the destruction of the mythocosm."* And what *is* that *"mythocosm"*? In essence it is simply the cosmos as Plato conceives of it, which consists, as we have seen, of two principal tiers: the *intelligible* and the *sensible*.[15] The crucial point however, let us recall, is that this division is not indicative of a separation or dichotomy, but that, on the contrary, it betokens a *relation*: that of a *sign* or signifier to its *referent* namely, as Borella affirms. Instead of cutting asunder or "breaking in two," it *connects*; that "connection," however, is not material—not "physical"—but *semantic*, properly so called. Which leads to the critical point of Platonist ontology: the fact, namely, that it is precisely by way of this *"semantic bond"*— connecting the sensible to the intelligible—that the former derives such reality as it has! In a word: *a sensible object exists by virtue of the intelligible archetype of which it is a kind of "image."*

This, in brief, is Jean Borella's explication of the "mythocosm": his reading, if you will, of Plato's myth. The "shadow" is *more* than simply an "effect"—more than the Hindu *"māya"* as popularly conceived. There exists a bond connecting the "shadow" to its source, which though itself "invisible," is in fact what enables us to "see." Such then is the *seeing* "inside the cave," St. Paul's *"as through a glass, darkly"* as distinguished from seeing "directly," described as *"face to face."*[16] And it might be worth noting that the aforesaid *"glass"* is in truth tantamount to the *"wall"* of Plato's cave.

Such was the "high" doctrine of the sapiential traditions, not only as transmitted in the Pythagorean, Platonist, and Neo-Platonist

---

15. The fact that we have divided the sensible into two domains—the corporeal and the intermediary—does not affect this paramount dichotomy; and I would add that the sapiential traditions offer many finer subdivisions of "the lower tier," based upon gradations of "sense." The Tantric science of *"cakras,"* for example, divides "the sensible" into five distinct levels, the lowest of which corresponds to our "sensible" realm. On this subject I refer to my chapter on *"Cakra* and Planet: O. M. Hinze's Discovery" in *Science & Myth* (Angelico Press, 2012).

16. 1 Cor. 13:12

schools, but in the corresponding Oriental schools as well. What I wish now to point out is that the Aristotelian philosophy represents a decisive break in this tradition, that Aristotle may in fact be seen as the first philosopher, at least in the West, to transition successfully from a *metaphysics* to a *physics*, howbeit of a distinctly premodern kind: a physics, let it be stressed, which has unquestionably its truth and its efficacy. The fact remains, however, that this Aristotelian doctrine constitutes already a decline: a shift *downwards* of the intellectual gaze, from the "vertical" orientation of Platonism to an inherently "horizontal" outlook, acceptive of corporeal entities at the level on which they are sensibly perceived. What in this—inherently hylomorphic—optic endows a corporeal entity with "being" is "form": *substantial* form, to use the Thomistic term, which though not subject to space, is perforce *subject to time.*[17]

The first decisive step in "the destruction of the mythocosm" has thus been accomplished by none other than Aristotle: the direction of the ontological outlook has been shifted by ninety degrees as it were—from "vertically upward" to "horizontal"—even as Raphael has depicted the scene.[18] But whereas it may seem that such a transition, if not indeed called for, is at least harmless, such is by no means the case. And no one, it seems, has expressed what actually stands at issue more prophetically than Nietzsche when he declared: "*We have abolished the true world. What has remained? The apparent one perhaps? Oh no! With the true world we have also abolished the apparent one.*" For so it proves to be: with the "abolition of the *true* world," the "*apparent* one" is abolished as well. Who could have imagined, even just a century ago, that the "destruction of the mythocosm"—which *is* in a sense "the true world"—would lead over the course of two millenia to the veritable "nothingness" opening up before us today in the physics of the twenty-first century!

But let us continue the story of the "descent." It needs to be noted next that the path to modern physics demands a second

17. It is to be noted that this situates *substantial form* on the intermediary plane.

18. I am referring to "The School of Athens" (1509-11).

pivotal step, to which the transition from the *intelligible* to the *sensible* opens the door: a shift, namely, from the *sensible* to the *geometric*, what Borella refers to as the "geometrization" of the real. When and where then, let us ask, did this second descent take place? Now clearly, it was *Galileo* who sparked that downward turn through the presumptive subjectification of what he termed "secondary" qualities:[19] for what remains—or *would* remain, better said—following the expulsion of the *bona fide* qualities are in fact *quantities*: i.e., attributes defined ultimately in terms of *spatial* bounds.[20]

The Galilean elimination of the qualities opens the door, thus, to a "geometrization" of corporeal reality, which from an Aristotelian point of view amounts to a spurious reduction of "content" to "container": of corporeal *substance*, that is, to its spatio-temporal bounds. It is hardly surprising, of course, that this "standing the world on its head," as we have termed this shift, cannot be accomplished overnight; and in point of fact, it has taken almost half a millennium of gargantuan endeavor for the end result to present itself upon the horizon in the guise of such marvels as the "quantum vacuum" or the "multiverse," wherein "being" has fully disappeared.

~

Modern physics came to birth with the publication of Sir Isaac Newton's *Principia* in 1687, almost half a century following the demise of Galileo Galilei. One can see in retrospect that the Newtonian physics—nowadays termed "classical"—constitutes the first of three phases, which held sway for a little over two centuries. What distinguishes this classical from post-classical physics categorically is the fact that it was based upon the post-Galilean notion of "matter," conceived as the space-occupying bearer of measurable attributes. Gone by now was not only the semantic link connecting the *sensible* to the *intelligible*—in other words, the

19. Mistakenly so, as we propose to show in chapter 5.
20. One must remember that time is measured by way of space.

"*mythocosm*"—but the Aristotelian *sensible* itself, conceived in terms of a substantial form. Not only, however, was this Galilean "matter" bereft of ontological foundation, but it lacked a rigorous empirical basis as well. Defined in effect as "mass extended in space," that renowned "matter"—touted by triumphant "materialists" as the substance out of which all things whatsoever are composed—turns out in truth to be no more than an imaginary "peg" for measurable attributes, such as mass and spatio-temporal coordinates.

The fact is that, beginning in the early decades of the twentieth century, a devastating epistemological critique began to eradicate the presumed conceptual foundations of that classical physics. As Sir Arthur Eddington points out, "previously scientists professed profound respect for the 'hard facts of observation'; but it had not occurred to them to ascertain what they were."[21] It was a time of radical transition: the "classical" foundation had in effect collapsed, and the subsequent phase—i.e., the quantum mechanical—was in process of formation.

Let us then consider that second phase of post-Galilean physics: the triumphant rise of quantum mechanics, associated forever with the illustrious names of Niels Bohr, Werner Heisenberg, and Erwin Schrödinger. To comprehend what stands at issue, one needs to recall that the classical notion of "matter" had all along been associated with the idea of atomism: the ancient Democritean notion that corporeal entities are made of minute indecomposable particles termed "atoms." And history confirms that this notion harbors a certain truth, which in fact came to light in 1897 with the discovery of the electron. For it turns out that corporeal entities can be disintegrated so as to give off what have subsequently been termed "quantum particles," which prove indeed to be "atomic" in the sense of being indivisible. To the amazement of the physics community, however, it happens that these indivisible entities prove not to be *bona fide* particles: that in fact, strictly speaking, *they do not actually exist!*[22]

---

21. From his Tarner Lectures, delivered at Cambridge University in 1938.
22. See, for instance, *Ancient Wisdom and Modern Misconceptions* (Angelico Press, 2013), ch. 1.

The preceding notion of "matter" had thus given way to what amounts to an insoluble conundrum for the physicist, faced as he was, henceforth, with the unpromising task of building what we perceive to be the corporeal world out of so-called "particles" which are not "actual" particles at all, but multilocating entities too weird to be so much as conceived in non-mathematical terms. What seems however to have disturbed the physics community more even than this ontological conundrum is the fact that the "geometrization" of the corporeal had not yet reached its term. Quantum particles—insubstantial to the point of "nonexistence" though they be—had not yet been reduced in effect to *a single unified mathematical structure*: and this is the perceived deficiency that has motivated leading physicists, ever since, to search for an "ultimate" physics, more basic even than quantum mechanics.

∾

Despite the fact that the presumed "content" of the universe had thus been dematerialized—reduced to so-called quantum particles which do not actually exist—there was still a formal separation between Content and Container in the form of "particles" which did not reduce to the terms of a pure geometry. In addition there were questions regarding the formation of atoms and so-called fundamental particles which quantum mechanics as such could not answer. And so—notwithstanding its almost miraculous success in predicting a vast array of physical phenomena, and the fact that it had given rise to a technology no one could have so much as imagined, it too has, in effect, become "yesterday's" physics. The fact is that hardly had the discovery of quantum mechanics been consummated than the physics elite shifted its gaze "downwards" once again: in its quest for a single unified theory which would "explain everything," the third and terminal phase in the evolution of post-Galilean physics had now begun in earnest.

The crucial fact to be noted is that this final step hinges upon the reduction of *time* to *space*: for so long as there is actual motion, there must be something that moves, which is to say that the aspired geometrization—the reduction of Content to Container—has not yet been achieved. And it is here that Albert Einstein enters upon the scene as the architect of the ultimate physics: for it is his conception of "space-time," precisely, that breaks the impasse and thus permits physics to enter its "promised land." Seemingly, that is! For it happens that space-time is a fake: there *is no* such thing. Indeed, one glance at the cosmic icon should suffice to make this clear: the fact that time precedes space ontologically implies that there is after all a globally defined simultaneity, in contradiction to Einstein's theory. From an ontological point of vantage the matter is in truth as simple as that!

But let us return to the ultimate physics: with his concept of space-time, Einstein has apparently succeeded in the requisite reduction of time to space. Representing time formally by a coordinate t in a four-dimensional space-time, Einstein has seemingly accomplished the impossible. The specifics, moreover, of this Einsteinian *Ansatz* are well known and need not detain us;[23] what concerns us is the fact that the way is now open to a formal reduction of Content to Container, which is to say that the ultimate *"geometrization" of the corporeal* appears now to be finally visible in the form of a *field theory* of some kind, a "quantum field theory" which one hopes will square with the empirical facts. But it appears at this juncture that Nature is no longer propitious to the physicist: the past half century—at the least—has been marked by persistent failure, which has apparently caused foundational theories to become progressively more abstruse and ever less testable. A point has been reached where the long-standing requirement of experimental verification, inclusive of "falsifiability," has been *de facto* abandoned; as Sabine Hossenfelder describes so well in her book,[24] that requirement has in effect been replaced by a norm of mathematical elegance, of "beauty" as she calls it.

The fact is that physics in its third or Einsteinian phase has ceased to be an authentically empirical science and is tending to morph into a *geometry*. Not only thus is the universe now conceived

---

23. A simple account is given in *Physics and Vertical Causation*, op. cit., pp. 48-51.

24. *Lost in Math: How Beauty Leads Physics Astray* (Basic Books, 2018).

in inherently geometric terms, but physics is becoming aprioristic in its *modus operandi*. It appears that Einsteinian symmetries are imposed and insisted upon, not because experimental results show this to be factual, but by fiat as it were. Meanwhile deviations between measured and predicted values are attaining record highs, most especially when it comes to quantities which can likewise be calculated on the basis of quantum mechanics. When Einsteinian physics encounters what appears to be the most accurate science known to man, discrepancies arise—up to 120 orders of magnitude no less![25]

〜

It is time to correct a statement made at the outset: the physicist has not in fact "stood the world on its head"; try as he may, he has not in fact succeeded in doing so. This incapacity, moreover, is by no means due to a lack of technical prowess: there is, quite obviously, no shortage of that—of genius even—within the higher scientific ranks. What is called for rather is a modicum, at least, of metaphysical discernment, sufficient to reveal the *a priori* impossibility of the contemplated project. The fact is that in pursuing the Galilean trajectory to its ultimate conclusion, physics is unavoidably approaching not simply a dead end, but ultimately a *reductio ad absurdum*: just think of so-called "many-worlds," or the unabashed absurdity of the multiverse! In truth it is the physicists themselves—and not the world at large—who are "standing on their head" by virtue of an upside-down worldview.

But whereas their *Weltanschauung* has been inverted since the days of Galileo and Descartes, physics itself remained a sound empirical science up to its third phase. In fact, by pursuing classical or so-called Newtonian physics to its natural conclusion—by shedding, step by step, what is in principle unmeasurable—physicists arrived finally at quantum mechanics, which, in my opinion at least, constitutes the true physics as well as "the end of the road." But be that as it may, it is in the third or Einsteinian stage that

---

25. As we shall see in the next chapter.

physics became fatally flawed through its relativistic "spatializa-tion" of time, which as we have noted, proves to be erroneous on metaphysical grounds. From this point onwards, it appears, foun-dational or "particle" physics ceased to be a viable empirical sci-ence, and began in effect to live on credit secured by discoveries achieved during its classical and quantum-mechanical phases.

～

What we have lost and desperately need to recover, strange to say, is nothing more esoteric than *the corporeal world* as it presents itself to every unimpaired human, inclusive doubtless of physicists. The problem moreover lies not in our sensory makeup, but in our pres-ent-day intellectual formation, which has become deviated to the point of schizophrenia. We absolutely need to rediscover that the grass is truly green, and that the song of birds resides not in the neurons. We need, in other words, to realize that what Galileo erroneously termed "secondary qualities"—such as color or sen-sible sound—pertain actually to the corporeal world, and nothing short of an ontological *metanoia*—a shift from the Galilean back to an Aristotelian or a Platonist outlook—will do.

But what is so important, one may wonder—for our under-standing of physics no less—in the "greenness of grass and the chirping of birds"? What stands at issue is the rediscovery of an ontological domain which physics—or more accurately, physi-cists—implicitly deny: the lower tier, namely, of what is tradi-tionally termed the *sensible* world: what we call the *corporeal*. The salient point is that *corporeal entities do not reduce to the categories of physics*: that they partake incurably of the Content as opposed to reducing to the Container. There is consequently a limit—an "upper bound" if you will—to what a mathematical physics as such can comprehend, which proves in fact to be far lower than one might expect—given that it excludes such ordinary things as green grass and chirping birds.

The implications of this realization for physics conceived as the foundational science are evidently drastic, and quite easy more-over to discern. One needs first of all to distinguish ontologically between the *quantum* realm and the *sensible* world—the *physical* as opposed to the *corporeal*, in our terminology—and note that these ontologically disparate realities coexist in the same spatio-temporal domain: it is thus the Content—rather than the Container—that proves to be split. We need then to remind ourselves what exactly physics is about; and I would maintain that no one has put it better than Lord Kelvin when he declared that *"physics is the science of measurement."* These words should be written over the entrance of every university physics department! It is easy enough to forget what they declare, and the moment we do, we have lost our way. So long as we do remember, on the other hand, we will never mistake the physical for the corporeal: for the physical—the quantum realm—is in essence *"the measurable,"* and *"that which measures"* is *perforce* the *corporeal.* I say *"perforce,"* because *"the measurable"* does not—cannot in fact—measure itself.

On the basis of Lord Kelvin's definition of physics as such one is led thus to the fundamental scenario of quantum mechanics as formulated in the Copenhagen interpretation, characterized by a categorical dichotomy between "the physical system" and the act of measurement that "collapses" the wave function. Our conclusion then is that *de jure* this scenario *holds universally for quantum physics at large.* The principle is simple, for it states no more than that *the measurable does not measure itself,* as we have noted.

The implications, on the other hand, of this "ontological axiom" are obviously immense: for it entails, after all, that any quantum theory which negates this categorical dichotomy of the measurable and the measuring—such as the Bohmian, for instance—is *ipso facto* invalidated. Or to put it another way: the principle implies that the Copenhagenist interpretation of quantum theory is in fact normative, the point being that the foundational dichotomy between the quantum system and "the environment"—i.e., the corporeal world—cannot legitimately be breached: the instant one does so, the discipline ceases to be physics, properly so called, and ceases consequently to be comprehensible. To which one should add that the history of particle physics from about the middle of the twentieth century onwards amply bears out this conclusion.

4

LOST IN MATH:
THE PARTICLE PHYSICS
QUANDARY

AROUND THE YEAR 1900, one can see in retrospect, a golden age of physics—the dazzling era of quantum mechanics—made its appearance; that golden age, however, proved to be short-lived. It began, if you will, in the year 1897 with J. J. Thomson's discovery of the electron, the first so-called quantum particle to see the light of day. And this epochal breakthrough was followed, in quick succession, by the detection of protons and neutrons: the structure of the chemical elements was about to present itself to scientific view in the form of the now familiar atom, with its heavy positively charged proton-neutron nucleus and light negatively charged electrons revolving around it in shells. At one stroke the mystery of the chemical elements, as enshrined in the periodic table, had now been resolved, reducing chemistry in principle to physics. Meanwhile, in the fateful year 1925, the dynamics of these so-called quantum particles was successfully encoded in the justly famous Schrödinger wave equation, which extends the classical laws of motion into the newly-discovered quantum realm. The venerable mechanics of Newton—which for three centuries had been perceived as the master key to the secrets of Nature—was now demoted to the status of an approximation, valid in the limit as the newly-discovered Planck constant h tends to zero.

In the wake of this mathematization of quantum theory, discovery succeeded discovery in its triumphant march of conquest.

The first enigma to be resolved was that of atomic spectra: those astounding sequences of spectral lines in the radiation emitted by atoms, which can be measured with uncanny precision and used to identify not only the chemical elements, but even their isotopes. It did not take long, following the discovery of the Schrödinger wave equation, for the likes of Wolfgang Pauli to calculate these spectral frequencies to degrees of accuracy never before attained. There was reason to believe that, in certain domains at least, physics had at last reached its ground level and was now conversant, not just with phenomena of various kinds, but with Nature's fundamental laws.

It was a period of unprecedented discovery, in which entirely new domains of inquiry and spheres of application sprang rapidly into view: think of the cyber revolution, to mention but one example, the foundations of which rest upon quantum theory. And let us not forget the atom bomb, which in its own way proclaims the stupendous magnitude of the quantum-mechanical breakthrough. What presently concerns us, however, are not the triumphs of quantum physics in the course of what we have termed its golden age, but the emergence rather of its limitations: for scarcely had the new theory come to birth, than phenomena came into view demonstrating its incompleteness.

The atomic nucleus, for example, posed problems the new physics alone could not resolve: what is the force, for example, which prevents the protons within an atomic nucleus from flying apart? Given that neither electromagnetic let alone gravitational forces are strong enough, it became apparent that there must exist as yet unknown forces. So too, soon enough, a host of other so-called quantum particles made their appearance: the positron, for example—a positively charged electron—was predicted in 1928 and observed three years later. And as a matter of fact, two other so-called "antiparticles"— the antiproton and antineutron—were likewise predicted in 1928 and observed in 1955 and 1956, respectively. Moreover new parameters descriptive of quantum particles—such as the measure of spin—came to be defined, and new particles, characterized in terms of these various indices, made their appearance in droves. What ensued was an embarrassment of riches, soon dubbed the particle zoo: a curious ensemble of weird semi-entities, with half-lives as small as 10^{-20} seconds, seemingly void of rhyme or reason, came into view. The need to bring order into this seeming chaos thus became paramount, and clearly the newly discovered quantum mechanics was not, by itself, up to the task.

By mid-century more or less, the marvelous new theory had apparently delivered its quota of epochal discoveries, and came to be viewed as something quite familiar and well explored. Yet when it comes to the foundations, Richard Feynman's "No one understands quantum theory"—so far from having been rendered obsolete—had rather been confirmed by the fact that all attempts to *understand* quantum theory—as opposed to merely applying it—seemed to indeed have failed. Nonetheless the theory continued to function magnificently wherever it was applied. Eventually, however, questions took center stage which seemed to call for other means, and thus the focus of fundamental physics began to shift: from quantum mechanics to the emerging vistas of particle physics. On its foundational level, physics was fast entering a subsequent epoch: an era defined by newly-formulated conceptions and staffed by a distinctly new breed of physicists.

By now—the second decade of the twenty-first century—the golden age of quantum mechanics has ceased, and both the style and the content of fundamental physics has changed substantially. One might say that we appear to know far more than in the preceding "golden" age, but with incomparably less assurance. Meanwhile it has become harder than ever for the non-specialist to catch so much as a glimpse of what is actually taking place on a foundational plane—to which one might add that the task of distinguishing scientific fact from pseudoscientific speculation has never been as daunting. What is urgently called for is a definitive treatise on the state of particle physics today, authored by an insider of the physics elite astute enough, honest enough—and above all, courageous enough—to tell the story as it is.

～

Enter Sabine Hossenfelder, a particle physicist of stature amply endowed with the aforesaid qualifications, who in a recent book, entitled *Lost in Math*,[1] has lifted the veil. She begins with a comment on the interplay between theory and experiment, noting that "in the last century, the division of labor between the theorists and the experimentalists worked very well. But my generation has been stunningly unsuccessful" (1).[2] And with astounding candor she admits: "I witnessed my profession slip into crisis... I'm no longer sure any more that what we do here, in the foundations of physics, is science" (2).

Between the glory days of quantum mechanics, moreover, and the present impasse, there is a transitional phase defined by a signal accomplishment: the theoretical formulation and subsequent empirical verification, namely, of the so-called standard model, which reduces the list of quantum particles—inclusive of the particle zoo—to precisely twenty-five "elementary" particles, of which all the rest are said to be composites. Out of the original electron, proton, neutron triad basic to chemistry, only the electron has survived as an indecomposable. I might mention that more than half a century had elapsed before the last of the standard model particles, the Higgs boson predicted in the 1960's, was detected: it took the Large Hadron Collider at CERN—boasting a tunnel circumference of 16 kilometers and a price tag of 6 billion dollars—to accomplish this feat.

Curiously enough, however, particle physicists, as a rule, are not pleased: no one, it seems, likes the standard model! For one thing, the newly found "elementary particles" have failed so far to contribute significantly to the understanding of what might be termed "our" world. Whereas the discovery of electrons, protons and neutrons has revolutionized science, beginning with chemistry, and has given rise to hitherto undreamed-of applications, the identification of semi-entities such as quarks and gluons has had no even remotely comparable effect beyond the confines of particle theory as such. It could in fact be argued that nothing of empirically appreciable consequence would be lost if we were to regard the erstwhile protons, electrons and neutrons as the indecomposable

---

1. New York: Basic Books, 2018.
2. Page numbers will be given in parentheses.

constituents of physical entities, so long of course as we bear in mind that they obey the laws of quantum theory as distinguished from Newtonian mechanics. As Sabine Hossenfelder explains: "This way we can speak of atoms and their shell structure, of molecules and their vibrational modes, and of metals and their super-conductivity—even though in the underlying theory there aren't any such things as atoms, metals, or their superconductivity, only elementary particles" (44). And as to the *magnum opus* of twentieth century particle physics, she concludes: "The most astounding fact about high-energy particle physics is that you can get away with knowing nothing about it."

It is unclear to what extent this "uselessness" has been an issue for present-day particle physicists. I will mention in passing that from the vantage point of the metaphysical traditions, their quest points "downwards" in reference to the *scala naturae*: from the pole of *morphe* to that of *hyle* namely, a descent which cannot but lead eventually to the "nothingness" of *prima materia*. And let me note that this view of the matter accords actually with Werner Heisenberg's stipulation regarding the ontological status of quantum particles as something "midway between being and nonbeing," which as such "are reminiscent of Aristotelian *potentiae.*" Ontologically speaking, it thus appears that contemporary particle physicists are actually moving in the wrong direction: "away from reality," towards the nether pole of *hyle*, where nothing at all exists. Their passion, according to this optic, is to hypostatize the ontological underside of whatever entities or semi-entities they encounter, as if "further down" signified "closer to reality." Is it any wonder, then, that physicists seem to be losing what Nick Herbert has so aptly called their "grip on reality"!

But hylomorphic considerations, evidently, do not carry much weight these days, and one might recall that even Heisenberg—the son of a classicist—stood alone among quantum theorists of the time in his ontological reflections. Meanwhile it should be noted that what to some extent at least compensates for the contemporary lack of ontological discernment is a passion for mathematics, coupled moreover with a deep and authentic apprehension of its beauty: the majesty, one could say, of its sovereign symmetries. And this is something Sabine Hossenfelder touches upon again and again, in full recognition of the well-nigh central role this factor currently plays in the dynamics of fundamental research. It appears moreover to be the underlying reason why no one among leading physicists is satisfied with the standard model, which seems nowadays to be perceived as a kind of makeshift, something that for aesthetic reasons alone could not possibly be the last word. As Sabine Hossenfelder explains:

> The standard model, despite its success, doesn't get much love from physicists. Michio Kaku calls it "ugly and contrived," Stephen Hawking says it's "ugly and ad hoc," Matt Strassler disparages it as "ugly and baroque," Brian Greene complains that the standard model is "too flexible," and Paul Davies thinks it "has the air of unfinished business" because "the tentative way it bundles together the electroweak and strong forces" is an "ugly feature." I have yet to find someone who actually likes the standard model. (70)

~

Yet its most serious flaw, in the eyes of the physics community, is doubtless the fact that it fails to account for one of the basic forces: gravity, that is. And perhaps the single most intensely pursued and erstwhile promising attempt to remedy this perceived deficiency is the discipline known as "string theory," which since about the mid-1980's has been the field of choice for many of the most highly gifted theorists. The sheer amplitude of its underlying mathematical structure—including spatial manifolds of up to fifteen dimensions—seemed to all but guarantee access to the ultimate secrets of the quantum realm.

A key condition which imposed itself from the start, termed "supersymmetry," entailed the fascinating prospect that every particle of the standard model had a "supersymmetry" partner, the observable parameters of which string theorists could in principle calculate. This enlargement, moreover—if it were to prove real—would "round out" the picture as described by the standard model, and hopefully remedy its proverbial "ugliness": the very "beauty" of supersymmetry seemed almost to guarantee this consummation. The expectation was rife, moreover, that at sufficiently high energies, nuclear collisions would produce these highly wished-for particles; and it is by no means implausible to suppose that it was to a large extent the lure of supersymmetry that motivated the construction of the Large Hadron Collider. But as Sabine informs us: "The LHC began its first run in 2008. Now it's December 2015 and no signs of supersymmetry have been found…" To which I would add that the situation remains unchanged to the present day, more than a decade later. Yet whereas many if not most theorists have by now abandoned the postulate of supersymmetry, others continue faithfully to believe the predicted "partners" will someday be found.

Meanwhile something far more portentous even than the apparent demise of supersymmetry has come to pass; as Sabine explains:

> In reaction, most string theorists discarded the idea that their theory would uniquely determine the laws of nature and instead embraced the multiverse, in which all the possible laws of nature are realized somewhere. They are now trying to construct a probability distribution for the multiverse according to which our universe would at least be likely. (173)

∾

"I begin to worry," she commented upon returning from a conference, "that physicists are about to discard the scientific method." And no wonder: for by the time one postulates a multiverse, the line has already been crossed. Let us be clear on this pivotal issue: it is absolutely impossible to accommodate multiverse theory to the scientific method due to the fact that no empirical communication between one universe and another can conceivably take place; no signal of any kind, after all, can transit from one universe to another. Inasmuch, therefore, as one can neither confirm nor falsify the existence of "another universe" by empirical means, the hypothesis of a multiverse is in principle bereft of scientific sanction. The so-called multiverse reduces thus to a kind of *deus ex machina* missioned to rescue a scientific theory which fails to square with the facts. How marvelous: an erroneous prediction comes in effect to be verified, howbeit in another world!

The multiverse, however, constitutes by no means the only incongruous conception currently in vogue: it appears that in this regard, contemporary astrophysics takes the prize. And I would add that since the confirmation of what has been dubbed "the axis of evil" based upon data from the Planck satellite, the fundamental premise underlying relativistic astrophysics—the so-called Copernican principle—has been demonstrably disqualified.[3]

Yet to this day particle physicists avail themselves freely of tenets pertaining to this now invalidated domain as if they were well-established scientific facts. A case in point pertains to so-called "inflation": "Inflation has some evidence speaking for it, though not overwhelmingly so" states Sabine, noting moreover that "physicists have further extrapolated this extrapolation to what is known as 'eternal inflation'" (104). Yet despite these uncertainties, "the string theory landscape conveniently merged with eternal inflation," she goes on to say (105). Nor does Sabine Hossenfelder stand alone in her doubts; as one of her colleagues remarked: "One can always find inflationary models to explain whatever phenomenon is represented by the flavor of the month... Inflationary cosmology, as we currently understand it, cannot be evaluated using the scientific method" (212). No wonder Sabine "is not sure anymore that what we do here, in the foundations of physics, is science"!

On a lighter note, she relates how a bump in the LHC data, observed on June 22, 2016, sparked the idea of a so-called "diphoton anomaly," which was subsequently put to rest by data gathered on August 4th: "In the eight months since its 'discovery' more than five hundred papers were written about a statistical fluctuation. Many of them were published in the field's top journals. The most popular ones have already been cited more than three

---

3. I have dealt with this issue in some detail in *Physics and Vertical Causation* (Angelico Press, 2019), chapter 5.

hundred times. If we learn anything from this, it's that current practice allows theoretical physicists to quickly invent hundreds of explanations for whatever data happens to be thrown at them" (235). From which she concludes:

> How long is too long to wait for a theory to be backed up by evidence...? Five hundred theories to explain a signal that wasn't and 193 models for the early universe are more than enough evidence that current quality standards are no longer useful to assess our theories. (236)

∾

The need of a theory "to be backed up by evidence": this takes us to the heart of perhaps the major problem which has increasingly bedeviled contemporary physics on its foundational plane. As physicists began to probe beyond the realm of such truly basic quantum particles as protons and electrons—and the gap between theory and application began to widen by leaps and bounds—"evidence" became ever more scarce. And by the time one arrives at the string theories—with their "invisible" dimensions—the temptation to dispense with that ever more inconvenient requirement became ever more imperious. Emboldened moreover by the fact that the pursuit of "mathematical elegance" had often in the past led to notable success, a growing cadre among the most gifted specialists succumbed evidently to the lure of "unconfirmed" and "unconfirmable" theory—right up to the *reductio ad absurdum* of the multiverse.

The most noteworthy commentary, in that regard, comes from the eminent physicist George Ellis, whom Sabine interviewed: "There are physicists now saying we don't have to test their ideas because they are such good ideas," he begins. "They're saying—explicitly or implicitly—that they want to weaken the requirement that theories have to be tested. To my mind, that's a step backwards by a thousand years" (213). Well said! The big question however, it seems to me, is whether "beyond a certain depth" empirical verification in the true sense is still feasible. Compare a proton, say, to a quark: one can still manipulate the former: place it in an electric field, for instance, and send it off in some specified direction. Despite "quantum indeterminacy," a proton is yet close enough to a "thing" to be somewhat observed and tested. So much so, in fact, that scientists have been tempted to "reify" such pseudo-entities in their imagination, "corporealize" them as it were. My point is that the picture changes drastically as one "descends" to a still deeper level of pure materiality: for as we have noted earlier, the ontological shift is directed towards *prima materia*, which technically speaking, does not actually exist. In place of a "theory of everything," therefore, the further evolution of fundamental physics appears thus to be moving inexorably towards a theory of "nothing at all": the very nothingness, if you will, of the multiverse.

Clearly Sabine Hossenfelder is onto something when she writes: "I can't believe what this once venerable profession has become. Theoretical physicists used to explain what was observed. Now they try to explain why they can't explain what was not observed" (108).

But regardless of whether contemporary particle physics may have "crossed the line" beyond which its *modus operandi* begins to fail, I would like now to touch upon another fundamental issue—utterly different—which strikes me as a likely source of invalidity. I am referring specifically to the wholesale acceptance of Einsteinian relativity, in both of its forms, along with the Einsteinian astrophysics, which has left its mark upon particle physics by way of pivotal conceptions such as "inflation" and "dark energy." We have already spoken of Einsteinian physics as definitive of its third stage, characterized by the reduction of time to space which completes the "geometrization" of the real.[4] We will now consider its wholesale incorporation into particle physics, which strikes me as uncritical, and perhaps the prime source of error in foundational physics at this time. I have contended elsewhere[5] in reference to

4. See ch. 3.
5. *Physics and Vertical Causation*, op. cit., ch. 5.

the special theory that Einstein's original argument was based ultimately upon ideological as opposed to scientific considerations,
and that crucial experiments have in fact proved him wrong. And
as regards the Einsteinian astrophysics, it appears that the theory
has been kept alive from the outset by way of *ad hoc* hypotheses,
missioned to neutralize the hostile data.

One of the earliest conceptions thus to impose itself upon fundamental physics is that of "inflation," which as we have come to
see, has risen to the status of a foundational premise in particle
physics. Inasmuch, however, as inflation alone could not fully
account for the expansion of the Einsteinian universe it was supposed to explain, a second *ad hoc* conception was brought into play:
i.e., that of "dark energy." I will leave it to the experts to inform
us whether this does or does not "explain" the seemingly acceler-
ated expansion of the Einsteinian cosmos at large. What I wish to
point out, first of all, is that the notion has sparked an immense
amount of research. As Sabine Hossenfelder informs us: "There
is an extensive literature on conjectured dark energy fields, like
chameleon fields, dilation fields, moduli, phantom fields, and quin-
tessence." The prime question, of course, is whether such a thing
as "dark energy" actually exists. And it might not be irrelevant to
point out that the concept bears the distinction of having led to
what Michio Kaku calls "the biggest mismatch between theory
and experiment in the history of science": to a vacuum energy,
namely, which is off by 120 orders of magnitude!

Leaving aside notions imported into particle physics from "big
bang" cosmology, the fact remains that particle physicists have,
from the start, built the basic conceptions of Einsteinian relativity
into the foundations of their discipline. The big question, of course,
is whether this "importation" can be justified on rigorous grounds.
To begin with, one might ask whether the reputed success of Richard Feynman's QED[6]—his quantum electrodynamics—may have
tilted the scales: for this would have been illegitimate, given that

---

6. Meanwhile facts have come to light which cast doubt on this reputed
success. See the remarkable article by the physicist Oliver Consa, "Something
is rotten in the state of QED" (https://vixra.org/pdf/2002.0011v1.pdf).

electrodynamics—in contrast to mechanics—has been in a sense "Einsteinian" from its inception by virtue of its Lorentz invariance. The question remains whether the wholesale imposition of Einsteinian symmetries upon which present-day particle physics is based is in fact justified; and for my part—speaking, to be sure, as an outsider—I surmise it is not.

It should be recalled, in this connection, that "foundational physics," as Sabine Hossenfelder points out, "is as far from experimental test as science can be while still being science" (27). The natural presumption that Einsteinian physics has been carefully and objectively "vetted" before being *de facto* elevated to the status of a foundational axiom proves to be in fact mistaken. From the start the new physics seems to have imposed itself upon the scientific community at large—not merely as a fascinating hypothesis—but rather as a long-awaited breakthrough, the "coming of age" almost of physics *per se*. As in the case of Darwinian evolution, evidence was hardly the prime concern: Einsteinian relativity was simply "too good" *not* to be true. The fact is that the Einsteinian *Ansatz* provides the key to the full geometrization of the universe—which to my mind is what ultimately explains how sober and brilliant physicists could have succumbed to a mere fantasy.

Be that as it may, particle physics has been Einsteinian from its inception. As regards the standard model, Sabine Hossenfelder informs us that "the gauge symmetries and the symmetries of special relativity dictate most of the standard model's structure" (55). What remains unaccounted for is the force of gravity, which needless to say, particle physicists have conceived from the outset in terms of general relativity. And that is where string theory comes into play. Although the theory originated as an attempt to understand strong nuclear interactions, its rise to fame began when, as Sabine points out, it was noticed that "the strings were exchanging a force that looked just like gravity" (172). In short, string theory "is a theory of quantum gravity" (188). And this is the theory, let us recall, which has given us the multiverse. But string theory too is now something in the past; as Keith Olive, an authority on the subject, confides to Sabine: "At the end of the day, the only thing we know to be true is the standard model" (65). It may be time to take a second look at the foundations.

～

It remains to ask why "in the field of quantum foundations, our colleagues want to improve a theory which has no shortcomings whatsoever" (5). Sabine is referring to that "golden age" of quantum theory with which we began these reflections: the good old physics of "protons, electrons and neutrons" which worked—and continues to work—so well. And she answers the question, at least in part: "It irks them"—the "foundational" physicists, namely—"that, as Richard Feynman, Niels Bohr, and other heroes of last century's physics complained, 'nobody understands quantum mechanics'" (6). Whatever may have sparked the movement to "improve upon" that physics, it turns out, had little to do with rectifying flaws or correcting deficiencies:

> Alas, all experiments have upheld the predictions of the not-under-standable theory from the last century. And the new theories? They are still untested speculations. (6)

Admittedly the physics "from the last century" had its limitations; but is this not perhaps a *sine qua non* of authentic science as such? One is reminded of Goethe's memorable dictum, "*In der Beschränkung zeigt sich der Meister*":[7] could it be that the prowess of science derives ultimately from its *Beschränkung*, its *limitation*? And does this not, finally, entail that the quest for a "theory of everything" is bound to fail? For my part, I am persuaded that such is indeed the case: in the final count, physics "works" precisely because it is *not* a "theory of everything."

What is it, however, which renders that—admittedly limited— "theory from the last century"—the one which "has no shortcomings whatsoever"—to be "not-understandable," to use Sabine's own term? It appears that Steven Weinberg—one of the

---

7. Literally, "In limitation the master shows himself."

greatest physicists of our time—has provided the crucial clue. In response to a question put to him by Sabine, he states:

> You can very well understand quantum mechanics in terms of an interaction of the system you're studying with an external environment which includes an observer. But this involves a quantum mechanical system interacting with a macroscopic system that produces the decoherence between different branches of the initial wave function. And where does that come from? That should be described also quantum mechanically. And, strictly speaking, within quantum mechanics itself there is no decoherence. (126)

The impasse of long standing could not have been stated more clearly! The reason why "no one understands quantum theory" resides thus in the measurement problem. And what renders this conundrum insoluble to the physicist is the fact that "strictly speaking, within quantum mechanics itself *there is no decoherence.*" Here we have it: the very *Beschränkung*, it turns out, which bestows upon the physicist his sovereign power to comprehend the physical universe, renders the measurement problem insoluble—to the physicist, that is!—by restricting his vision to the realm of the physical as such.

What is it, then, that this vision excludes? I answered this question in the first paragraph of my first book: "It excludes the blueness of the sky and the roar of breaking waves" I wrote, "the fragrance of flowers and all the innumerable qualities that lend color, charm and meaning to our terrestrial and cosmic environment." To which of course the "scientific" response will be: "But these are all *subjective* attributes: that color and that sound—that's all in your head!" Here we have it: the *Beschränkung* is yet in force! It has not been transcended: the aficionados of physical science have apparently become *de facto* incapable of transcending it.

Let me say apodictically that there *is in truth no scientific basis* whatsoever for the aforesaid response, the aforesaid reductionism; and in truth *there cannot be.* What confronts us here is not a scientific, but indeed a *philosophical* issue. The subjectification of the qualities is actually none other than the Cartesian postulate, what Alfred North Whitehead terms *bifurcation*, which affirms that reality divides ontologically into two disjoint realms: an external world consisting of *res extensae* or "extended things," all else being relegated to a subjective domain of so-called *res cogitantes* or "thinking entities." Introduced in the seventeenth century by René Descartes,[8] it has ushered in the period known as the Enlightenment, and has bedeviled our intellectuals ever since. My point, now, is that this premise proves to be none other than the *Beschränkung* upon which physics *per se* has been based since the days of Newton.

No wonder "nobody understands quantum theory"! The very *limitation* upon which physics as such is based renders this impossible. But whereas our *Weltanschauung* has been askew since the Enlightenment, what has changed with the advent of quantum theory is that henceforth *physics itself can no longer be understood on a bifurcationist basis.* The true significance of the measurement problem resides in the seemingly paradoxical fact that decoherence does take place, even though "within quantum mechanics itself there is no decoherence." But if not "within," there must be a "without": and this recognition opens the door to a rediscovery of the authentic "unfiltered" world, along with the fullness of our humanity.

The remedy, therefore, is simple in the extreme: rectify this misconception—expunge the fallacy of bifurcation, root it out—and the resolution of the "insoluble" measurement problem stares you in the face: at one stroke quantum theory ceases to be "not-understandable," and turns comprehensible in a trice.[9]

---

8. Strictly speaking, the idea traces back to the atomism of Democritus, his celebrated fragment: "*By convention there exist color, the sweet and the bitter, but in reality only atoms and the void.*" And let me note that both Plato and Aristotle were wise enough to recognize this doctrine as untenable.

9. A summary presentation of that resolution may be found in my lecture, "From Schrödinger's Cat to Thomistic Ontology," reprinted in *Ancient Wisdom and Modern Misconceptions* (Angelico Press, 2013), chapter 1. A complete and rigorously "non-bifurcationist" account of physics per se is given in *The Quantum Enigma* (Angelico Press, 2005).

Which brings us finally to the question: how does this resolution of the long-standing quantum enigma impact the present-day particle physics quandary? For my part, I entertain the hope that this rediscovery of the integral cosmos will open the way out of that seeming impasse.

5

DO WE PERCEIVE THE
CORPOREAL WORLD?

FROM TIME IMMEMORIAL mankind has presumed that we do look out upon the world: that the objects we perceive are not in fact "mental," but "real." One understands, of course, that it is possible to err: that a rope in semi-darkness may be mistaken for a snake; yet the very circumstance that perception can thus miscarry seems to confirm that, normally, it does not. It appears, moreover, that right up to the period known as the Enlightenment, this realist view of visual perception was espoused, not only by the simple and untutored, but by even the most eminent philosophers as well. There have been dissenting voices, to be sure: the case of Democritus, for instance, who claims that "*color, the sweet and the bitter*" exist in appearance alone, "*but in reality, only atoms and the void.*" Yet this view was definitively rejected by the major philosophic schools of antiquity.

Since the advent of the so-called Enlightenment, however, this heterodox *Weltanschauung* has gained well-nigh universal acceptance within the educated strata of Western civilization. The shift began early in the seventeenth century with Galileo's categorical distinction between what he termed *primary* and *secondary* qualities, the first of which were said to pertain to the object itself—which once again was thus conceived to consist of "*atoms and the void*"—whilst the latter, such as color, were assigned to the human observer, who allegedly projected these "mental" qualities upon the external world. And this Galilean reinvention of Democritean atomism came soon to be championed by René Descartes, who elaborated and systematized that philosophy, thereby preparing the ground for the birth of modern physics, an event which came to pass in the year 1687 with the publication of Sir Isaac Newton's *Principia*.

Two points need to be emphasized: first, that from the time of Newton to the present day, every Western scientist—virtually to a man—has more or less unknowingly based his worldview upon this Cartesian dichotomy; and second, that there exists not a shred of scientific evidence in support of that Cartesian claim. What is more: in the course of the twentieth century some of the leading philosophers—headed by Alfred North Whitehead—have declared the "bifurcationist" premise to be misconceived, yet manifestly to no avail: our official and reputedly "scientific" *Weltanschauung* remains to this day Cartesian to the core.

With the rise of neuroscience, moreover, word is out that the Cartesian view of visual perception, say, has now been confirmed on empirical grounds. Sir Francis Crick for instance—of "double helix" fame—promulgates this reputed fact: "What we see," he informs us, "appears to be located outside, although the neurons that do the seeing are inside the head."[1] One may, of course, wonder by what empirical means this remarkable fact has been ascertained—but that is a question Sir Francis does not address: he evidently deems it sufficient to affirm—in the name of neuroscience no less—that it *is* "the neurons" that see.

It has come to pass, however, that this claim has been seriously challenged by a neuroscientist named M. R. Bennett, in collaboration with an Oxford philosopher named P. M. S. Hacker, who give the following response:

> This is at best misleading, since in perception the brain learns nothing—it is the person whose brain it is that perceives and learns something in perceiving. What we see does not *appear* to be located outside us. What we see is *necessarily* located outside our body... The neurons without which we would not see anything are located in our heads, but, of course, it is not those neurons that see

1. *The Astonishing Hypothesis* (Simon & Schuster, 1995), p. 104.

or that "do" the seeing... There is no sense in which "the world" is in anyone's head...[2]

These clear and eminently sensible observations will no doubt be welcomed by many as a breath of fresh air. Yet granting the validity of Bennett and Hacker's realist stance, the question remains how that prodigy of visual perception is in truth accomplished. It is of course to be understood that the neuronal structures in the visual cortex of the brain do have a necessary role to play, as Bennett and Hacker presume; yet it is far from clear at this point, what precisely that role might be.

Crick himself in effect admits as much when he tells us that "we can see how the brain takes the picture apart, but we do not yet see how it puts it together,"[3] a statement based upon two assumptions, both of which prove to be false: for the first so-called "picture"—the one the brain is said to "take apart"—is in fact the so-called retinal image, which is actually not a "picture" at all, whereas the second—the one the brain is unable to produce—turns out to be the mountain or the tree we happen to be looking at.

But whereas Crick is evidently mistaken regarding what it is that we perceive, he is right in pointing out that the brain itself cannot get us there. What confronts us in this impasse, first of all, is the ontological difficulty of making one thing—whether it be a mere image or an external object—out of many: out of the on/off states, in this instance, of countless neurons in the visual cortex of the brain. This is the famous "binding problem" which has engaged top-level neuroscientists for decades, with no resolution in sight.

I would point out, first of all, that basing themselves apparently upon a realist view of visual perception, Bennett and Hacker seem to dismiss this pressing question as a pseudo-problem:

For the color, shape, location and movement of the blue delphiniums swaying in the breeze cannot be taken apart (there is no such thing as separating these attributes from the objects of which they

---

2. *Philosophical Foundations of Neuroscience* (Blackwell, 2003), pp. 127-8.
3. Op. cit., p. 159.

are attributes), and the color, shape, location and movement of the delphiniums cannot be brought together in the brain, since these attributes are not to be found in the brain, either together or separately.[4]

Yet regardless of whether it makes sense to speak of "separating the color from delphiniums," doubts arise when they maintain that "it is precisely this confusion [between the blue delphiniums swaying in the breeze and whatever may be found in the brain] that informs the neuroscientists' characterization of the 'binding problem'." For whereas Bennett and Hacker are unquestionably right in pointing out that "this confusion" has profoundly affected the neuroscience community, it would be a serious mistake to dismiss the issue *per se* as a pseudo-problem. For even after the aforesaid misinterpretation has been diagnosed and rectified, the question remains how a neuronal mechanism can play a role in the visual perception of the external world.

There is thus, most assuredly, a *bona fide* "binding problem"; and though that problem may generally be misconceived by neuroscientists, it proves to be nonetheless profound and in fact intractable under Cartesian auspices. Even from the perspective of Bennett and Hacker, moreover, there must, in the final count, be a unifying principle to connect the multitude of neurons—inclusive of their on/off positions—to the act of perceiving, failing which the entire neuronal mechanism would evidently be irrelevant to visual perception. Or to put it another way: if indeed the brain acts as a computer—which hardly anyone disputes—there must be something or someone, in effect, to "read" that computer: to bring into unity what, on the level of the physical brain, is fragmented into a myriad microscopic components. But that "principle" is something our scientists are *de facto* unable to conceive, for it happens to be the very conception excluded from our *Weltanschauung* by the Cartesian *Ansatz* itself. I wish however to point out that the restoration of that crucial component—its reintegration into our worldview—can in truth be effected with ease: one need look no

4. Op. cit., p. 139.

further than the ontology of St. Thomas Aquinas, which refers to that element as a *"substantial form."*

~

This fact should moreover come as no surprise: after all, the substantial form of an animate being is none other than what is traditionally termed a *soul*;[5] and there was a time, not so very long ago, when every child knew that each of us has one. But getting back to the binding problem: it is evidently the soul that "binds" not only the neurons in the visual cortex, but every discernible component of the body into the unity of the living organism. Is it any wonder that—having jettisoned this crucial component—we should face a "binding problem"!

The question arises, of course, how the soul or substantial form itself can accomplish that seemingly miraculous feat of "uniting" an unspeakably vast multitude of entities—be they neurons, "atomic particles," or what you will—into a single living organism; and let me emphasize that the formal re-introduction of substantial forms in itself hardly resolves the mystery. To grasp what actually stands at issue it is evidently needful to have in addition some insight into the nature and resultant powers of that soul. And to this end we need to avail ourselves of the premodern—pre-Cartesian—cosmological wisdom of mankind, beginning with the tripartite nature of the human microcosm, answering to the designations *corpus*, *anima*, and *spiritus*.[6] What is it, then, that distinguishes these three ontological domains? As we have seen,[7] it proves to be the bounds of space and time: for whereas the corporeal domain is evidently subject to both space and time, the intermediary—sometimes termed the "astral" plane—is subject to time alone, while the

---

5. A "rational soul" in the case of a human being: for it happens that the human soul differs categorically from the souls of animals (not to speak of plants, which own what is Thomistically termed a "vegetative" soul).

6. See *Physics and Vertical Causation* (Angelico Press, 2019), especially chs. 6 and 8. The ontological distinction between soul and spirit (*psyche* and *pneuma*) is to be found explicitly in St. Paul, e.g., 1 Thes. 5:23.

7. See ch. 2.

spiritual is subject to neither. The essential point, now, regarding "souls" is that *they pertain to the intermediary domain*, which is to say that they are subject to time *but not to space*. And here at last we have the "missing" piece of the puzzle: for it is by virtue of being "supra-spatial," precisely, that the soul has the capacity to "bind" into one substance, one living organism, a multitude of spatially separated components—millions of neurons, for instance, in the visual cortex of the brain. No wonder the neuroscientists are baffled; for it happens that the binding problem *is in fact insoluble on the corporeal plane*.

Getting back to Bennett and Hacker: we fully concur that perception does not take place on the level of neurons, and that "the brain does not 'construct a perceived world', but enables the animal to see a visible scene." And we concur likewise when they go on to say:

> Since seeing a tree is not seeing an internal picture of a tree, the brain does not have to construct any such picture. It merely has to be functioning normally so that we are able to see clearly and distinctly. It does not have to take a picture apart, since neither the visual scene nor the light array falling on the retinae are pictures. It does not have to put a picture back together again, since what it enables us to do is to see a tree (not a picture of a tree) in the garden (not in the brain).

It could hardly have been put more clearly! My point, however, is that these observations do nothing to resolve the binding problem, but rather bring to light its seeming insolubility insofar as they expose the ontological chasm that separates the firing of neurons from the seeing of a tree.

It is possible, of course, to ignore the problem: to simply let neurons be neurons and trees be trees. Yet the impasse remains, and the philosopher, more profoundly even than the neuroscientist, will have failed in his mission unless he discovers the element or principle connecting the percipient to the object perceived. And that is where metaphysics—like it or not—enters into the discourse. What is called for, as I have stated, is the time-honored conception of the *soul* as the substantial form of the living organism, along with the realization that this soul—traditionally termed *psyche* or *anima*—unlike the components it binds, is not subject to the divisions of space. It is the resultant "transcendence of spatial separation" on the part of the living organism that resolves the binding problem.

∽

Having ascertained that we do actually perceive a tree in the garden as distinguished from a mere image of the tree, we arrive finally at the decisive question: how does neuronal activity enter into the process of visual perception? What role, precisely, do neurons play in that process: why, in fact, are they needed at all? Now, I know of only one cognitive psychologist who has successfully answered this question on empirical grounds—and oddly enough, Bennett and Hacker do not so much as mention his name. I am referring to James J. Gibson, who began his career at Cornell University in the early 1940's on a government research grant missioned to ascertain how one "sees" the aiming point of a complex motion, and discovered to his amazement that the information contained in the so-called "retinal image" does not in fact suffice to determine that point. Then and there he recognized what his colleagues have apparently failed to grasp to this day: i.e., that *this fact alone disproves the premise upon which the prevailing "image theory" of visual perception is based.* At one stroke Gibson realized that every domain of research pertaining to visual image psychology—for instance, the massive endeavor to explain how the so-called "third dimension" is supposedly constructed from retinal data—is doomed from the outset to fail.[8] In a word, Gibson realized that the theory of visual perception stands in need—in *desperate* need—of a new paradigm.

Following this breakthrough Gibson set off on a course of empirical investigation in quest of a paradigm for visual perception that stands the test of empirical scrutiny. And some three decades

8. This, of course, is not to say that visual image theory gets nothing right at all: it works perfectly well, for instance, in the prescription of spectacles.

later, this hard-headed no-nonsense empiricist averse to scientistic fairytales did discover what he was searching for, and proceeded to formulate what he termed the "ecological theory of visual perception" to emphasize that what is normally perceived resides neither in the brain nor in the consciousness of the percipient, nor indeed in the world as conceived by the physicist, but in the so-called *environment*, which is thus tantamount to what I term the *corporeal* world. What we perceive proves to be a mountain or a tree in the garden, as humanity had in fact believed from time immemorial.

It would take us far beyond the scope of this chapter to present so much as a summary account of Gibson's revolutionary theory of visual perception.[9] I wish rather to reflect upon the astounding fact that this "ecological" theory has actually broken the Cartesian stranglehold and re-established realism—in fact what might legitimately be termed "naïve realism"—on rigorous empirical grounds. Gone is the twin fantasy of *res cogitantes* and *res extensae*! The redness of a rose is not after all a "secondary" quality as Galileo had imagined: a mental apparition mistakenly projected upon a colorless external entity. Following four centuries of utter confusion, it turns out that roses are indeed red, and that the reputed "simpletons" were wiser in fact than the pundits of the Enlightenment, who to this day are living in a fantasy world—or *would be*, better said, if they consistently believed what they affirm.

Suffice it to say that the crux of Gibson's discovery resides in the recognition that visual perception is based, not on a so-called retinal image—which actually does not exist—but on what he terms *invariants* in the ambient optic array, which prove to be inherently *forms*. The fact is that forms give access to the external world, to "the blue delphiniums swaying in the breeze": for as Aristotle observed, it happens that the very being of material entities consists—not of what is nowadays termed "matter"—but yes, of *forms*. It is *only*, therefore, by way of forms that the perception of

9. The best reference, to be sure, is Gibson's classic monograph, *The Ecological Approach to Visual Perception* (Lawrence Erlbaum Associates, 1986). For an overview I refer to "The Enigma of Visual Perception" in *Science & Myth* (Angelico Press, 2012).

an external object can actually take place. Bravo, James J. Gibson: you deserve ten Nobel Prizes at the very least!

The great question, now, is how that Gibsonian "pickup of invariants" is actually accomplished, and how, moreover, the complexities unearthed by contemporary neuroscience fit into the emerging picture. The first thing to be noted in that regard is that the "computer" paradigm is simply not up to the task: admittedly it may explain this or that secondary phenomenon, but when it comes to the central issue—how we perceive, not a mere image of whatever kind, but actual corporeal entities—on that score it gives us not so much as a clue. A neuronal computer, like any other, is as one knows equipped to handle what electrical engineers term Shannon information, and as such it does presumably have a role to play; yet by no stretch of the imagination can a computer of any kind—even one "made of meat"—pick up or transmit Gibsonian invariants: *color* for instance.

It turns out, moreover, that the pickup of invariants from the ambient optic array requires movement, demands in fact a search: and this alone disqualifies the erstwhile "camera" paradigm of visual perception. For whereas a camera can produce a succession of static images, that is not what we actually perceive. What we perceive is the result of a search which involves much more than the eye conceived as a camera. The requisite receptors must, for example, include the muscles responsible for those exceedingly rapid eye-movements known as "microsaccades," which play a decisive role in that "pickup of invariants." The actual "instrument of visual perception" comprises thus a substantial part of the human anatomy in addition to the eyes and visual cortex. Moreover, this "instrument" cannot, strictly speaking, be conceived in *mechanistic* or Cartesian terms. Gibson himself does not perhaps bring this out as clearly as one might wish: he is after all an empiricist, not a metaphysician. Yet even so it emerges from his findings that *the actual instrument of visual perception does not in fact reduce to "the sum of its parts,"* which is to say that the Gibsonian pickup does not reduce to a mechanical process.[10] Hard

10. I will mention in passing that this in itself identifies that pickup of invariants as a *vertical* act.

as it may be for the post-Enlightenment mind to conceive, we are dealing actually, not with a robot, but with a living organism! The perceptual system, so far from functioning as a mechanism, is in fact animated by the substantial form of the living organism: the very form traditionally termed a *soul*. Not only, thus, is that soul needed to "bind" the neurons in the visual cortex as noted before, but it is in fact required to "bind" the entire organism, and in so doing render it both living and sentient.

It needs hardly to be pointed out that Gibson's theory of visual perception proves to be epochal in its significance and implications—which oddly enough appears to be the reason why only a comparative few have so much as heard his name: for one can see at a glance that the Gibsonian breakthrough suffices *de jure* to bring down our post-Enlightenment *Weltanschauung* at a single stroke. The fact is that in refuting the Cartesian epistemology, Gibson has disproved the fundamental premise upon which our supposedly "scientific" worldview rests to this day. We must not expect to see statues of him erected any time soon.

∼

The fact is that *we do visually perceive the corporeal realm*. Inasmuch, then, as there are thus two ways of knowing the external world— visual perception plus the *modus operandi* of physics—there are in a sense "two" of everything corporeal: "two tables" for example, the corporeal and the molecular let us say. The question now presents itself *which of the two is "real."*

It needs first of all to be understood that this question is incurably philosophical—metaphysical in fact—which is to say that it cannot be resolved by the methods of physics. There is moreover an obvious asymmetry between these two ways of knowing, inasmuch as the *modus operandi* of physics hinges upon sense perception, whereas the converse does not hold. This fact connects with the Thomistic metaphysics, which conceives of sense perception as the basis of human knowing.

Let us then, in conclusion, examine the issue at hand in light of Thomistic philosophy. What is it then, Thomistically speaking, that differentiates the corporeal domain—the world as known through sense perception—from the physical? The first thing a Thomist is prone to note is that physical entities are defined in exclusively quantitative terms, which is to say that the physical universe constitutes an inherently *quantitative* realm. To the Scholastic mind, this recalls the dictum "*numerus stat ex parte materiae*": "number derives from the side of matter." The allusion, clearly, is to hylomorphism, which conceives of corporeal entities as consisting of two ontological components: *morphe* or form, plus *materia* or matter, conceived as pure potency. The point is that neither *materia* nor *forma* exist by themselves: it "takes two to exist." The fact, therefore, that "number derives from the side of *materia*" entails that *forma* is missing in the physical realm: it turns out that *an ontological domain defined in purely quantitative terms is perforce sub-existential.* And here, in brief, we have *the resolution of the quantum enigma*: a quantum particle, say, by virtue of its purely "quantitative" nature, is consequently situated "midway between being and nonbeing," as Heisenberg has put it, and can be in truth no more than an Aristotelian *potentia*. What "actualizes" such *potentiae*, moreover, is none other than the act of measurement, and it does so by contributing the missing ingredient: namely *form*. What on the other hand distinguishes the quantum realm itself from *materia* as such—termed *prima materia*—is the fact that it is endowed with a *quantitative* structure. The physical domain reduces thus to what the Scholastics termed a *materia secunda* or "secondary" *materia*. It is interesting to note, finally, that St. Thomas speaks of a *materia quantitate signata*: a *materia secunda*, that is, which he characterizes as "marked by quantity." For my part, I cannot conceive of a more perfect metaphysical characterization of the physical universe.

What I wish now to convey is that these Thomistic considerations not only enable us to understand the ontology of the quantum world, but that they provide the key to a metaphysical comprehension of physics as such. They establish, first of all, *the necessity of quantum mechanics as a rectification of Newtonian physics* by the fact that the *res extensae* upon which Newtonian physics is based *do not actually exist*. It was consequently foreseeable from the outset that as the operative conceptions of Newtonian physics sharpened and became ever more operational under the influence of such philosophical schools as logical positivism, a point would be reached at which the postulated *res extensae* dissolve, as it were, into the mist of the quantum realm. It becomes manifest that *quantum mechanics is simply physics divested of its Cartesian illusions*—simply, *physics come into its own.*[11] The transition from classical to quantum physics was therefore predictable on metaphysical and indeed Scholastic grounds, based upon the *modus operandi* of physics as such.

It may be well to note at this juncture that a metaphysics is always in play: our choice lies merely between truth and error. And as we have seen, the Newtonian edifice had a metaphysics of its own—the Cartesian, namely—the hegemony of which, as we have noted as well, persists to the present day, notwithstanding the fact that this philosophy flies in the face of quantum theory. One may ask oneself thus: how is it possible to espouse a metaphysics at odds with a physics we know to be true? But whereas this is not possible so long as one understands that physics, it *is* quite readily if one doesn't. It turns out that Richard Feynman was right: no one *can* in fact "understand quantum theory" so long as he remains a Cartesian.

What is called for is a radically different metaphysics; and it will no doubt come as a surprise to many that the requisite doctrine is readily to be found in a culture we have maligned with gusto since the Enlightenment. It appears that the shoe is now on the other foot: as we have come to see, even a cursory comprehension of Scholastic metaphysics suffices to resolve the impasse which has stymied our top intellectuals since the advent of quantum theory.

---

11. The reader may recall that we have made that claim in chapter 3.

6

PONDERING BOHMIAN MECHANICS

As if the bewilderment over quantum theory were not formidable enough in itself, the Bohmian *Ansatz*—which purportedly returns us almost to classical clarity—has added what may prove to be the most perplexing element of all to the mix.

Let us start at the beginning by considering what exactly the "hidden variables" account of quantum theory is meant to accomplish. What motivated David Bohm to radically restructure the mathematical formalism of what might well be termed the most accurate physics the world has ever seen? The answer may in fact be gleaned from the preface to his last book,[1] and is actually manifest in the subtitle itself, which reads: "An *Ontological* Interpretation of Quantum Theory."[2] So far from being a positivist, or an aficionado of some kindred school, it appears that Bohm was actually a metaphysician at heart, and that it is this propensity—so very rare among scientists in our day—that sets him apart. What he had his eye upon was not primarily *how things appear*, but what in truth they *are*. And therein, it seems, lies the motivation behind his "hidden variables" approach, which was designed—neither to simplify, nor to extend the range of quantum mechanics—but rather to manifest its *ontological* significance. As his co-author B. J. Hiley explains:

1. D. Bohm & B. J. Hiley, *The Undivided Universe* (Routledge, 1993). Bohm himself died unexpectedly "just as the final touches were being put to the manuscript."
2. Italics mine.

Indeed the most radical view to emerge from our deliberations was the concept of wholeness, a notion in which a system formed a totality whose overall behavior was richer than could be obtained from the sum of its parts. In the ontological theory that we present here, this wholeness is made manifest through the notion of nonlocality, a notion that is seemingly denied by relativity.[3]

I find these observations profoundly significant. There are here three assertions: (1) that the authentic system constitutes a whole which is *more than the sum of its parts*; (2) that it is precisely *this wholeness* that gives rise to the phenomenon of nonlocality; and (3) that the question remains whether Einsteinian relativity and nonlocality are actually incompatible, as appears indeed to be the case. Leaving aside the third—with which we need not concern ourselves inasmuch as we have rejected the Einsteinian theory—let me note, first of all, that I concur wholeheartedly with the first two affirmations, and actually perceive in them a key to the ontological comprehension of quantum mechanics.

What then, in Bohm's view, has gone awry in the standard[4] formulation of quantum mechanics that motivated him to revise the theory from the ground up? The error resides, supposedly, in the omission of an essential element from the quantum-mechanical description: *the measuring apparatus* that is. The "wholeness" of the measuring process, Bohm insists, has thus been compromised: what has been jettisoned—what the standard wave function fails to describe—stands in fact at the very center of what physics is about: "Indeed, without the measuring instruments in which the predicted results appear, the equations of the quantum theory would be just pure mathematics that would have no physical meaning at all."[5] True enough: physics is after all "the science of measurement" as Lord Kelvin observed, which is to say that it is in the act of measurement, precisely, that physics comes into its own. Now, for Bohm this

3. Op. cit., p. xi.
4. Quantum theory, as formulated by its founders, beginning with Werner Heisenberg in 1925.
5. Op. cit., p. 2.

means apparently that the formalism of physics—its actual equations—must somehow incorporate the measuring process itself. It is hardly surprising, therefore, that he perceives the standard formulation of quantum mechanics—which conceives of measurement simply as a "collapse" of the wave function, a discontinuity which cannot be bridged by any differential equation—to be no more than a means of prediction which sheds no light on the underlying process. And, clearly, this omission of what he takes to be *ontologically* essential constitutes, for Bohm, the fundamental defect he wishes to correct through an extension and refinement of the mathematical formalism. It was not mere mathematical bravado, thus, but rather a profound metaphysical striving that led Bohm to his astounding breakthrough, and explains why he regarded the most accurate physics the world had ever seen to be woefully deficient. As Bohm and Hiley observe dismissively: "All that is clear about the [standard] quantum theory is that it contains an algorithm for computing the probabilities of experimental results."[6]

According to the dominant strains in the philosophy of science—such as logical positivism—this is of course precisely what a quantum mechanics is meant to do. What primarily interests Bohm, on the other hand, as we have seen, is not the prediction of measurable effects, but *the ontology of the cosmos*, its very *wholeness*. As Bohm and Hiley explain: "We have chosen as the subtitle of our book 'An Ontological Interpretation of Quantum Theory' because it gives the clearest and most accurate description of what the book is about": of what in fact Bohmian mechanics as such is about. "The question of determinism is therefore a secondary one, while the primary question is whether we can have an adequate conception of the reality of a quantum system."[7] And this brings us back to the issue of *measurement*: that is where the mystery resides.

And this explains the radical departure of Bohmian mechanics from standard quantum theory: the fact that the "hidden variables"

6. Ibid., p. 1.

7. Ibid., p. 2. It is to be noted that in the standard formulation, the quantum system as such is ultimately conceived as something "less than real": as in a way "reminiscent of Aristotelian *potentiae*," to quote Heisenberg.

approach claims to bring the process of measurement itself into the quantum-mechanical description. In place of a two-tier process consisting of a "Schrödinger evolution" described by the wave equation, followed by a discontinuous and mathematically unpredictable "projection" effected by an act of measurement, Bohm conceives of a continuous evolution of the combined system—the measurable plus the measuring—in which all is governed by a single system of differential equations, without "inexplicable" discontinuities.

To a non-Bohmian quantum physicist this sounds miraculous, and it is safe to say that prior to 1952 when David Bohm apparently accomplished this feat, not a single quantum theorist of repute would have thought it could be done. In light of quantum mechanics as envisaged in its standard formulation, the ultimate physical components were conceived to be "half particle, half wave," which is to say that they could in fact be neither. The notion that there exist *bona fide* particles, moving under the guidance of a so-called "pilot wave" in well-defined orbits calculable by way of differential equations descriptive of the integral system—the measurable plus the measuring—seemed thus to have been rigorously excluded once and for all. And as a matter of historical fact, when the notion of "hidden variables" was first enunciated by Louis de Broglie at the 1927 Solvay Conference, it was quickly disposed of, eviscerated in fact by the presiding heavyweights.

Yet the founding premise of de Broglie-Bohm physics is simple in the extreme, and by no means unattractive to physicists: it affirms that *real* particles interact with a *real* wave to define *real* trajectories determined by a system of differential equations—the trick is only to discover these equations, to write them down and apply the resultant formalism to the solution of actual problems. And to the amazement of just about every quantum physicist—including possibly Bohm himself—it turns out that this can apparently be done, and that the resultant mechanics yields in fact the same results as the standard theory.[8]

8. To which one should add: at least most of the time. The matter is currently in dispute among quantum theorists.

As could have been predicted, the reaction to Bohm's epochal discovery has been mixed. On the opposing side I would, above all, cite Werner Heisenberg, the first to write down—in a form of his own—the famous equation definitive of quantum mechanics, which in a thousand experiments has never yet yielded a false result. In *Physics and Philosophy*, published six years after Bohm's breakthrough, he gives a summary critique of "hidden variables" theory, which at the very least is dismissive of Bohm's ontological claims. Heisenberg's central objection is that "the hidden parameters of Bohm's interpretation can never occur in the description of real processes if quantum theory remains unchanged,"[9] in support of which he refers to physical scenarios on an atomic scale, such as those underlying the Compton effect. In a word, Heisenberg perceives the "hidden variables" of Bohmian mechanics as inherently speculative conceptions which prove to be *de facto* meaningless on the level of physics properly so called, which he evidently conceives as an interplay of theory and experiment. In this optic Bohmian mechanics reduces to an unnecessary and potentially misleading means to arrive at results which can be obtained far more directly by way of the standard theory.

To be sure, Bohm perceives the matter differently: as a "metaphysician at heart," he perceives it from a different point of vantage. What interests him primarily is the ontology, the underlying "wholeness" of which the measurable properties are effects, and one almost senses a note of sarcasm when he laments that, in the standard or so-called Copenhagen interpretation, all that remains is "an algorithm for computing the probabilities of experimental results."

〜

It is that Bohmian conception of "wholeness" thus—the notion of a "totality whose overall behavior is richer than could be obtained from the sum of its parts"—that calls for reflection. I too am persuaded that authentic ontology demands such a "wholeness," and

9. *Physics and Philosophy* (Harper & Row, 1958), p. 132.

concur that this entails—and in a way explains—the phenomenon of nonlocality: an apparently instantaneous action at a distance known to occur in the quantum world. What interests me above all, however, in the aforesaid declaration is something Bohm and his followers seem not to perceive: an implication, namely, discernable by way of Aristotle's conception of "quantity" as that which "admits mutually external parts." For it entails a recognition the significance of which could hardly be overstated: the fact, namely, that *a whole which does not reduce to "the sum of its parts" cannot be fully described in quantitative terms.*

One more step is needed to reveal in full what has thus come to light: for if such be the case, what else, in addition to quantities, does that wholeness comprise? The question is *ontological*: what else, besides *quantities*, can there be? And here the Aristotelian categories prove to hold the key: the complement of *quantities*, namely, are *qualities*, which unlike the former, do *not* "admit mutually external parts," and cannot therefore be quantified.[10]

We arrive thus—by way of rudimentary conceptions tracing back over two thousand years—at a recognition which has become virtually unthinkable in our day: i.e., that inasmuch as the cosmos constitutes an "unbroken wholeness," it does not reduce to quantity, and cannot therefore be fully described or comprehended in terms of physics alone. No formulation of physics therefore—not even Bohmian mechanics—can give us more than a partial or limited comprehension of the universe inasmuch as it leaves out of account the qualitative dimension of reality. And one might add, in regard to this "qualitative dimension," that in light of traditional ontology this proves actually to be the "higher" of the two inasmuch as it is expressive of morphe or form as opposed to hyle or materia. To be precise, the "quantities" to which physics would

10. The notion that the color red, for example, reduces to a light frequency is hardly worthy of response. Red—or redness, if you prefer—is a *quality*, and anyone who takes it to be a frequency is simply confused. That there is a correlation between color and light frequency cannot of course be denied, but to say that the two are one and the same is to speak nonsense. Aristotle had it right: there is a category difference between *quantities* and *qualities*; and this fact proves to be critical to an understanding of physics.

reduce the universe correspond in fact to numerus, which according to Scholastic doctrine "*stat ex parte materiae*": stems from the side of materia. As we have noted from the outset, the purview of physics is severely restricted.

⟨∿⟩

To the neutral observer it cannot but seem astonishing that even the most brilliant physicists—including those who espouse "unbroken wholeness"—have apparently failed to grasp that the world does not in fact consist of quantities: it makes one realize how powerful a force the *Zeitgeist* actually proves to be. Since the onslaught of the Enlightenment in the seventeenth century, Western intellectuals have been hypnotized, as it were, by the Cartesian dichotomy—the notion that reality splits neatly into an "objective" world of *res extensae*, definable in purely quantitative terms, complemented by a "subjective" realm of so-called *res cogitantes*, entities which have no existence in the "objectively real" world. We have come thus to regard the external world as a realm bereft of qualities, which can consequently be described without residue in purely mathematical terms. It appears that Descartes has cut asunder what in the act of knowing is conjoined: we have forgotten what it is "*to know*."

It thus behooves us to reflect anew upon the "act of knowing" that connects us to the "external" world, beginning with visual perception, upon which that knowing is in fact primarily based. I would point out, first of all, that the subjectivist interpretation of visual perception is not—and in fact, *cannot be*—based on empirical grounds: no amount of experimentation can establish that what we "actually perceive" when we look at an external object is simply "in our head." The point is that *perception* is something radically different from *sensation*. To be sure, the current theory of visual perception, espoused almost universally by cognitive psychologists, is indeed subjectivist to the core; it needs however to be realized that this subjectivism is actually implied by the presiding "retinal image" or "camera" paradigm itself, which has dominated the discipline since the Enlightenment. As we have however come to see in the preceding chapter, it happens that James J. Gibson has demonstrated the invalidity of visual image psychology and re-established the fact that we do perceive the corporeal world.

As should by now be apparent, we have broached the subject of perception precisely because *measurement* is consummated in such an act. If it be true, therefore, that *colors*, for instance, prove *not* to be mere *res cogitantes*, this would imply that the instrument of measurement does not in fact reduce to a *res extensa*. And this in turn would of course entail that standard quantum mechanics is right after all in distinguishing categorically between the physical system and the measuring apparatus, and that, by the same token, Bohmian mechanics is mistaken in doing otherwise. My point is that Gibson's empirically based claim that we actually perceive what he terms the "environment"—instead of a "visual image"—militates against the Bohmian theory.

∾

It follows that the Copenhagen formulation of quantum theory—the one that, according to Bohm, reduces supposedly to a mere algorithm for computing probabilities—had it right. For if indeed it be the case that what we perceive in the act of visual perception is *not* "in our head," but pertains actually to the "external" domain, this enables us—*forces* us, in fact—to distinguish ontologically between the physical system, characterized by quantitative attributes, and the measuring instrument, which consequently owns *qualitative* attributes, such as color, as well. What confronts us in the measuring configuration is a meeting of two distinct ontological planes—what I term the *physical* and the *corporeal*—resulting in the actualization of a physical state which constitutes indeed a transition from *potency* to *act*, as Heisenberg maintains. One arrives thus quite naturally at an *ontological* understanding of quantum mechanics.[11]

11. As first enunciated in *The Quantum Enigma* (republished by Angelico Press in 2005). For a summary account I refer to my monograph, *Physics and Vertical Causation* (Angelico Press, 2019). A brief introduction may also be found in my article, "From Schrödinger's Cat to Thomistic Ontology," republished in *Ancient Wisdom and Modern Misconceptions* (Angelico Press, 2013), chapter 1.

What I find objectionable in "hidden variables" theory is its very structure: by incorporating the measuring apparatus, namely, into the physical system, it renders invisible the ontological dichotomy upon which quantum mechanics as such is based. It thus implicitly negates the true ontology, *the very thing Bohmian mechanics was meant to discern*, and instead perpetuates the myth of Democritus: the misbegotten notion that qualities—*"color, the sweet and the bitter"*—constitute illusory superimpositions, and that at bottom reality reduces to *"atoms and the void."* Inasmuch, therefore, as the distinction between the quantum system and the measuring apparatus proves to be metaphysical, it turns out that by his very *Ansatz*, Bohm has closed the door to an ontological comprehension of quantum mechanics. I find it tragic that the commendable endeavor to bring to light the ontology of quantum physics should have been based upon a denial of the very distinction upon which that ontology rests.

That failure to distinguish ontologically between the physical system and the instrument of measurement likewise impedes a second major recognition: the discernment, namely, of a hitherto unknown mode of causation, which I refer to as *vertical.*[12] What distinguishes this mode is the fact that it does not operate "in time" by way of a causal chain, and cannot therefore be expressed in terms of differential equations. It is chiefly in fact by way of an ontological discontinuity that such an "instantaneous" mode of causation can be discerned.

∾

It behooves us, finally, to reflect somewhat upon "the mystery of knowing" as such: the relation, that is, of a *knower* to the *known*. If we say that knowledge constitutes a kind of union, what is it, then, that connects the two poles? And if that union proves indeed to be "a certain oneness" as Aristotle claims, how is it possible for two things thus to "become one"? I see no other way to deal with this question effectively than that of hylomorphism. By distinguishing

---

12. See *Physics and Vertical Causation*, op. cit.

between what the Greeks termed *hyle* and *morphe*, and the Romans *materia* and *forma*, conceived to be related as "recipient" to that which is "received"—somewhat as clay is receptive to the likeness of Socrates—it becomes possible for a second "receptor" to receive the very same *form*.[13] For inasmuch as a *form* has no substantial being of its own—is not itself a "thing"—it is not separated, spatially or otherwise, from a potential knower. It is possible then, for a subject with a "*materia*-like" receptivity, to "receive" that *form*, even as clay can receive "the form" of whatever it be. And so—"in a certain sense" as Aristotle has it—the knower and the known "become one."

Returning to the case of visual perception, it thus appears that the connecting link uniting the percipient to the perceived can indeed be none other than *form*. Since the onset of the Enlightenment, on the other hand, we have been led to believe that the link could only be a physical process, which however entails that in truth there exists no "link" and indeed no "objective" knowing. What we perceive is then a phantom-like "image" of some kind, which would mean that we do not actually perceive the "external" world at all. Predictably so: because as we have noted before, nothing other than *form* permits the miracle of *knowing*![14] It should come as no surprise, therefore, that whereas no trace of *forms* is to be found in the "visual image" theory of perception, *forms* do in fact play the key role in Gibson's theory, which hinges upon a so-called "pickup of invariants from the ambient optic array"—which is to say that the "invariants" prove in fact to be *forms*.[15] It is thus indeed by way of *forms* that the "environment"—corresponding to what I term the *corporeal* domain—can be visually perceived.

13. The italics are meant to indicate that the word is used in the specific sense of *forma*.

14. One might add that the reason resides in the fact that *to know* is something *sui generis*: is not, in other words, the same as *to be*; and this crucial distinction was clearly discerned by philosophers, from Plato to Alfred North Whitehead.

15. It should be noted that Gibson himself does not make this identification, and it is unclear to what extent he was philosophically inclined. He comes across as a "hard headed" empiricist, howbeit with powers of discernment rarely to be found.

~

Getting back to David Bohm, I wish now to show that in the course of his ontological reflections he actually came within a hair's breadth of making essentially the same discovery as James Gibson: namely that visual perception hinges upon a pickup of what turns out in the end to be—yes, a *form*. Yet whereas at first glance this may seem incongruous, it is scarcely surprising in view of Bohm's fascination with the idea of "wholeness": the "invariants" by which we perceive what Gibson terms the "environment," and which are said to be given at every location within that ambient optic array, exemplify after all the idea of "the part containing the whole" at which Bohm arrived by way of quantum theory.

Let us recall two quantum-mechanical findings, in particular, which point to that conception: first, the phenomenon of non-locality—an inalienable characteristic of quantum theory—which affirms that particles half a world distant may be so connected as not in truth to be separated at all, an effect which cannot in truth be understood in mechanistic terms, and may consequently be recognized as an effect of vertical causation; and secondly, that according to quantum theory "the whole may actually organize the parts, not merely through the strong connection of very distant elements, but also because *the state of the whole is such that it organizes parts.*"[16] Gone is the key principle of classical physics: the idea that the universe can be understood in terms of its "atomic" parts. Not only, thus, does the whole fail to reduce to the sum of its parts, but it has actually the capacity to act upon the parts, to "organize" them.

But even this proves ultimately to be insufficient: from these initial reflections Bohm appears to have advanced to the full-blown realization that—in some exceedingly recondite sense—*the whole is actually present to or "contained in" every part.* Thus, alluding to a holograph, Bohm notes that "each part is an image of the whole object," which leads him to the pivotal conception of what

16. David Bohm, *Unfolding Meaning* (Ark Paperbacks, 1987), p. 7; italics mine.

he terms *enfoldment*: "Therefore every part contains information about the whole object . . . information about the whole is *enfolded* in each part."

My point is that it was this crucial insight that enabled Bohm to rediscover, in essence, the salient feature of Gibson's theory: this "*enfoldment*," namely, *proves to be the key to the enigma of visual perception*. Bohm tells us so himself with the utmost clarity:

> The light from all parts of the room contains information about the whole room and, in a way, enfolds it in this tiny region going through the pupil of your eye, and is unfolded by the lens, and the nervous system—the brain—and somehow consciousness produces a sense of the whole room unfolded in a way which we don't really understand. But the entire room is enfolded in each part. This is crucial, because otherwise we wouldn't be able to understand what the room was—the fact is that there is a whole room, and we see the room from each part.[17]

Now this is simply brilliant, and tantamount in fact to Gibson's notion of "invariants in the ambient optic array": Bohm too, in his own way, has in effect rediscovered "forms." Their respective accounts of visual perception concur, moreover, right up to the point where the light enters "the pupil of your eye," following which they become antithetical. The fact is that "inside the head" Bohm's description is based upon the "camera" paradigm, in keeping with what the "experts" have been telling us for the past so many centuries; and on this question Gibson obviously holds the advantage. Having found "what the experts say" to be fallacious, and having arrived by way of an exhaustive empirical inquiry at a new paradigm, he envisages visual perception in a radically different way. The point, however, is that this new approach proves to be none other, inherently, than what Bohm himself was in fact seeking: for it happens that the Gibsonian "pickup of invariants from the ambient optic array," so far from reducing to a mechanical process, hinges precisely upon the idea—yes, of *enfoldment*!

17. Ibid., p. 10.

Gibson's account of visual perception is thus in essence "Bohmian," and would doubtless have delighted the master himself.

❧

Finally, let us not fail to note that the rediscovery of realism by way of the Bohm-Gibson "enfoldment" theory of visual perception confirms what we have deduced directly from Aristotle's categories: the fact, namely, that so far from reducing to the world as conceived by the physicist, the universe proves to be incomparably richer and more sublime than we have been taught to believe. It turns out that our terrestrial environment, in particular, is endowed with a plethora of *forms*, which not only combine with *materia* to produce corporeal entities, but enable us also to know these corporeal entities by way of perception. Yet ever since the Enlightenment, Western civilization has warred against *form* in a Promethean endeavor *to reduce the world to quantity*. And whereas the universe as such evidently remains what it is, our conception thereof has morphed to the point that a realist ontology has become *de facto* unthinkable. As Pavel Florensky—the venerable and pansophic priest executed in 1937 by a Soviet firing squad—observed a century ago: "On the basis of positivism and materialism and, in general, of trends of thought that reject the essential reality of *forms*, there is no place for realism."

7

ASTROLOGY:
THE SCIENCE OF
WHOLENESS

IT NEEDS FIRST of all to be understood that the "wholeness" to which we allude is none other than the "tripartite wholeness" previously defined,[1] conceived in light of Platonist ontology. As we have seen, it can be depicted iconically in the form of a circle in which the center refers to the primary realm, not subject to the bounds of time or space, and the circumference represents what we know as the corporeal world. As for the interior of that iconic circle, it is indicative of a long-forgotten domain subject to time alone, which we refer to as the intermediary domain. What primarily confronts us in this tripartite ontology is thus a dichotomy between the supra-temporal and the temporal realms, which is tantamount to the Platonist distinction between the *intelligible* and the *sensible* orders, followed by a division of the sensible into a *gross* or "corporeal" and a *subtle* or "psychic" stratum. The central claim of Platonism—which I deem to be definitive—can now be stated with astounding brevity: it affirms that *the sensible derives its reality from the intelligible*, to which it stands in principle as a *signifier* to its *referent*.[2] And that ontological recognition, I contend,

---

1. See ch. 2.
2. This interpretation has been pioneered by the French Platonist Jean Borella. See for instance *The Crisis of Religious Symbolism* (Angelico Press, 2018), Part I.

proves to be the master key to the enigma of astrology.[3]

To imagine that stars and planets—*as currently conceived*—affect human character and destiny is doubtless an absurdity. Yet whatever the present-day aficionados of astrology may think in that regard, this is *not* what astrology actually affirms. It needs first of all to be understood that if the cosmos at large were, even remotely, what we nowadays take it to be—a conglomerate, namely, of astrophysical entities scattered throughout the immensities of space, made up of quantum particles—if *that* were in truth the case, astrology would indeed reduce to the "exploded superstition" it is generally taken to be. The point, however, is that the cosmos *does not in truth reduce to the conceptions of our current astrophysics*: if not even the simplest corporeal entity—a pebble, say, in the palm of my hand—reduces to an aggregate of quantum particles, what to speak of the cosmos at large!

The key to the enigma of astrology, I have said, resides in the Platonist ontology: nothing less than that high—and perhaps, in a sense, "esoteric"—teaching will suffice. It informs us that *the reality of all things pertains ultimately to the intelligible realm*— which entails that so long as stars and planets are conceived as spatio-temporal entities, they can ultimately be no more than *signs* pointing to an *intelligible referent*. Such causality, therefore, as enters into the purview of astrology, originates—*not* in stellar or planetary masses moving through space—but precisely in that supra-temporal sphere Platonists refer to as the *intelligible* world. That causality, therefore, is not effected by a temporal transmission through space—the kind known to physics, to which I refer as *horizontal*—but proves to be what I call *vertical* causation: a kind that does *not* operate "in time."[4] The planets, as they sweep out the zodiac with its twelve constellations, may thus be likened to the

---

3. The fact that the sensible is divided into two domains—the corporeal and the intermediary—has no direct bearing on this issue.

4. Unlike the "horizontal" modes of causation which pertain to the domain of physics, *vertical* causality is not mediated by a temporal process, but acts instantaneously. See *Physics and Vertical Causation* (Angelico Press, 2019), which defines and justifies VC in the context of quantum mechanics.

hands of a cosmic clock announcing the "seasons"[5] of the sensible world, which in truth they no more "cause" than the ticking of a clock can cause flowers to blossom and trees to bear fruit.

～

Not only however is the cosmos tripartite in its integrality, but according to traditional doctrine, every ontological stratum reflects that tripartition on its own plane. What then constitutes the resultant tripartition of the corporeal world? It consists of the *Earth* as representing the corporeal, the *planetary realm* the intermediary, and the *stellar* the intelligible world. The spatio-temporal world proves thus to be an *iconic* representation of the tripartite cosmos: howbeit an *inverted* one, inasmuch as it places the Earth—representative of the lowest stratum—at the center. The fact is that the corporeal world—*Platonically conceived*—constitutes a natural icon of the cosmos at large: the *geocentric* icon, one may call it. And *it is upon this icon that astrology is based.*

It should however be recalled that the "stars and planets" as thus conceived do not by any means reduce to stars and planets as described by the physicist, which are subcorporeal. Nor are these "stars and planets" strictly speaking perceptible: what we actually perceive when we look at a star or planet is after all no more than a spot of light which we take to be the star or planet.[6] The aforesaid division of the corporeal world into its terrestrial, planetary, and stellar realms is not "merely symbolic" as one tends nowadays to imagine, but *real* in a profoundly metaphysical sense. The corporeal world proves to be *ontologically tripartite* because it manifests—in a distinctly Platonist sense—the cosmic trichotomy itself. It *must* in fact do so by virtue of what it *is*: i.e., an "image" of the tripartite whole. And this ontological fact is vital: it is the reason, ultimately, why astrology is *not* in truth "a superstition"!

5. Gen. 1:14

6. The Christian reader may recall that St. Paul distinguishes ontologically between terrestrial and celestial bodies, for instance in 1 Corinthians 15. On this issue I refer to *Ancient Wisdom and Modern Misconceptions* (Angelico Press, 2013), chapter 6.

But let me recall that this image is "inverted": that the Earth, though central, holds in truth the ontologically *lowest* rank. And I will note in passing that this geocentric icon constitutes actually the basis of the Ptolemaic cosmography, which has never been "disproved" but merely discarded in post-Galilean times, when its ontological significance was no longer understood.[7]

This brings us to a third "whole" crucial to astrology: the one upon which in fact it is technically based. I am referring to what is termed a *horoscope*: one may think of it as a symbolic diagram representing the geocentric icon with its terrestrial center plus planetary and stellar spheres, as these present themselves to an observer at a particular time and terrestrial location. What counts is the position of the seven astrological planets relative to each other, and to the twelve zodiacal signs. We shall not enter into the casting or reading of horoscopes, a subject well covered in any number of creditable sources. Our objective is rather to indicate, in the broadest strokes possible, the *ontological basis* upon which astrological science rests. To this end we first need however to complete the picture; for it happens that we have so far left out an additional "wholeness" which enters crucially into play: namely, *our own*.

The decisive—and long-forgotten—fact is that in the integrality of their being, *every man, woman, and child constitutes a veritable microcosm*: a tripartite whole, consisting of *corpus*, *anima*, and *spiritus*, which moreover is in a sense *"isomorphic"* to the macrocosm. This means that whatever astrological elements define the global structure of the universe—thereby rendering it a *cosmos*—have their counterpart in the microcosm: namely *in us*. Unbelievable as this surely appears to be from our "atomizing" point of view, whatsoever is definitive of the astrological cosmos—the planets Venus and Jupiter, for instance—has its counterpart in the *anthropos*, the microcosm that we are. Of course, from a contemporary

7. On the subject of geocentrism I refer to *Physics and Vertical Causation*, op. cit., pp. 48-62.

point of vantage, nothing could be more absurd! Suffice it to point out, however, that from a *Platonist* point of view, just the opposite holds true: the central fact that *being* derives from the intelligible order entails that every authentic wholeness *cannot* but replicate—*in its own inimitable way*—the very wholeness of the cosmos as it pre-exists on the intelligible plane.

The reason why astrology is *not* a "superstition" resides finally in the well-nigh unbelievable fact that we carry the "stars and planets" within our integral being: to think of them as so many light-years distant—*that* proves in the end to be the real superstition! The problem, however, is that only in terms of an inherently Platonist ontology can this be grasped: the Aristotelian already falls short in that regard—what to speak of the post-Galilean ontologies, which have in effect eliminated wholeness as such, and ended up with a quantum dust that turns out, *in fine finali*, not to exist.[8]

The crucial point of the Platonist worldview resides in the fundamental dichotomy between the *intelligible* and the *sensible* realms: between a kind of being *not subject to the conditions of time and space*, and the time-conditioned modes terminating in the corporeal. What needs above all to be grasped is that *corporeal* entities derive such being as they have—*not*, most assuredly, from quantum particles—but from an *intelligible referent*, in relation to which they serve as a *sign*: which is to say that in truth their reality is ultimately *semantic*, to use Jean Borella's term. And it is upon this *semanticity*, precisely, that the *modus operandi* of astrology is based. The reason, thus, why stars and planets can serve as *signs* is that—in the final count—*they are* precisely that.

∽

The tripartite constitution of the human microcosm testifies to its aeviternity, its supra-temporal nature; and hard as it may be for the contemporary mind to accept, the time and place of our birth, so far from being "accidental," is actually indicative of "who it is" that has, here and now, entered the corporeal world—which is in

8. See ch. 3.

fact *precisely* what the natal horoscope discloses in its own way. We have all become persuaded that there exists a stringent causation based upon the interaction of minute corporeal—or subcorpo- real—parts, yet we balk generally at the very notion of a causality emanating from wholeness. And even when it has been proved that a mode of causation based upon parts does not suffice—can- not account, for instance, for the production of so-called "complex specified information"[9]—we continue as a rule to deny that a cau- sality based upon *wholeness* is so much as conceivable. Yet the fact remains that this "inconceivable causality of wholeness" turns out to be *the only kind that permits intelligence, morality, art*—and in fact, *science itself—to exist.*

Getting back to the natal horoscope: what else can it be than a representation—in the "language" of astrology to be sure—of the person in the wholeness of his being, as identified by the time and place of his birth? And that very "language" is precious and irreplaceable inasmuch as it is not contrived—not "invented"—but is based upon the "true morphology" of man: his authentic "struc- ture" as the microcosm he is. One might go so far as to claim that one requires the language of astrology to perceive a person in the actuality of his being, which lies far deeper than his "physical" and "psychological" characteristics.

Astrology deals with wholeness, we have said; and it can do so because its *modus operandi* is itself *holistic* as opposed to analytical. Of course there are rules—for beginners or amateurs, one might say—to the effect that "Jupiter in Gemini," say, is indicative of this or that; and no doubt dicta of that kind have their place. But they do not add up to what may rightly be termed "the science of astrol- ogy"—any more than textbook rules concerning scales, chords, and meters add up to a "hearing" of a sonata or a symphony. The point, once again, is that music—all authentic art for that matter—deals with *wholes* which in fact *do not reduce to the sum of their parts.* And this fact is crucial: for otherwise one is dealing—not in truth with

---

9. I am referring to a mathematical theorem proved by William Dembski in 1998. I have given a readable introduction in chapter 9 of *Ancient Wisdom and Modern Misconceptions*, op. cit.

a whole—but with a mere aggregate, which of course is precisely what our physical sciences are about: the name of the game is to *decompose*, to eliminate in effect whatever authentic "wholeness" may initially exist.

But whereas these contemporary sciences prove thus to be inherently reductive—destructive, that is, of authentic wholeness—such is by no means a *sine qua non* of science *per se*, which is to say that there do in fact exist *bona fide* sciences of a "non-reductive" kind. The point, however, is that these differ fundamentally from the reductive genre in their *modus operandi*, and bear moreover a *qualitative* significance over and above their quantitative content. It may come as a surprise that even pure mathematics admits such non-reductive modes, and that in fact the earliest schools were concerned mainly with the qualitative aspects of arithmetical and geometric facts. Such is the case especially in the Pythagorean tradition, where a profound connection between mathematics and the arts—beginning with music and architecture—comes into view. That connection persisted in fact right up to the advent of modern times, which is to say that the contemporary reduction of mathematics to an abstract formalism—and ultimately, to set theory—constitutes actually a "post-Enlightenment" phenomenon.[10] Before that time a "number" was an integer or a ratio of such, which to this day is termed a "rational" number, and these numbers— so far from reducing to pure "quantity"—possessed a qualitative identity: a *wholeness* namely, of which the mathematician of our day has no longer the ghost of an idea. To catch a glimpse of what stands at issue in these long-forgotten mathematical sciences one may recall that, in the context of musical harmony, it is precisely a halftone that takes you from a major to a minor key: from the dynamic "world of the Sun," if you will, to the pensive "sphere of the Moon," its qualitative complement—a fact concerning which our contemporary understanding of mathematics can obviously tell us nothing at all.

10. This process of reduction is epitomized in Whitehead and Russell's monumental *Principia Mathematica*, published in three volumes between 1910 and 1913.

≈

What I am driving at is that the same holds true for the natal horoscope, which has to do likewise with qualities and wholeness as opposed to quantities and parts. To be precise, it represents the person in the wholeness of his being as mirrored in the corresponding aspect of the cosmos, defined by the positions of the planets with respect to each other and the twelve zodiacal signs corresponding to the time and place of his birth. And this makes sense because a human being is himself a *microcosm*: a replica "in miniature" of the integral cosmos one might say. The natal horoscope may thus be viewed as a "description" of a particular man or woman in terms based upon a universal wholeness which has been studied and codified by astrologers for thousands of years. And whereas this may seem irrational in a culture conditioned to view wholes as nothing more than an aggregate of parts, the picture changes drastically once it is realized that the matter stands in truth just the other way round: that it is actually *the whole that gives rise to its parts.*[11] It then makes perfect sense moreover to suppose that the wholeness of the macrocosm contains within itself the virtually infinite ensemble of potential microcosmic wholes, as represented by attainable horoscopes.

What this entails is that the panorama of human life is not simply haphazard—void of rhyme or reason as one might say—but comports with a principle of order. But again, what stands at issue is not an order based upon "parts"—not a Laplacian determinism to be sure—but an order based upon wholeness such as we encounter, for instance, in the spheres of music and art. The natal horoscope may thus be compared to the key of a musical composition that imposes a recognizable characteristic, which however enables rather than impedes freedom of expression. It is in truth that very wholeness expressed by the natal horoscope which enables creative activity as such—beginning, one might say, with the production of what Bill Dembski terms "complex specified information," which

---

11. Even the slightest knowledge of biology should suffice *de jure* to make this clear.

of course is merely the quantitative imprint of what is thereby produced. In short, what we are and are able to accomplish as a human being is encoded in the natal horoscope. And let us note that the "reading" of that horoscope cannot ultimately be reduced to rules. Rules are there, of course, for the beginner, and give one a start; but they are there to be transcended. The ultimate reading of a horoscope is like the "hearing" of a symphony: in the final count, wholeness speaks to wholeness alone.

Let me note once more, however, that to the contemporary mind—which has all but forfeited the very idea of wholeness—all this has become incomprehensible, and is consequently taken to be absurd. All such a mind is able to discern—be it in the cosmos at large or in a human being—is an accidental assemblage of parts, precariously held together by the action of blind forces; and we can all agree that if such were the case, astrology would in truth reduce to an absurdity. It happens however that such is *not* the case.

We have referred to astrology as "the science of wholeness": it needs however to be understood that what interests the astrologer is that wholeness—in both its macro- and microcosmic manifestations—as distinguished from the causality connecting the two. We add nothing to astrological science by pointing out that it hinges upon vertical modes of causality; we only explain why the Encyclopedia Britannica refers to it as "an exploded superstition." It is not the concern of the astrologer to understand the causal connection between a human microcosm and the cosmic macrocosm: his task is to *read* a horoscope, not to explain why it works. And this "reading" is as much an art as it is a science. The novice in astrology, to be sure, labors to apply its textbook "rules" as best he can;[12] yet a master "reads" a horoscope very much the way a true musician

---

12. I do not wish to demean these "rules," or suggest that they do not apply: amazingly enough, they do apply to an astounding degree! My point is rather that the "wholeness" of which the horoscope is a representation can in fact be accessed fully only by way of an intellective act: a perception of wholeness rather than the application of rules. This holds true, incidentally, even in the purely quantitative sciences of our day: in pure mathematics itself namely, as we learn from Gödel's justly famous theorem. For a short introduction and gist of the proof, see *Physics and Vertical Causation*, op. cit., pp. 41-4.

"reads" a score: he "hears" the horoscope, one is tempted to say.

Philosophically speaking, the fact that astrology is based exclusively upon vertical causation entails that it can only be understood in Platonist terms. The causal nexus between the microcosm and the macrocosm, as we have noted before, proves to be the kind that connects a *sensible* signifier to an *intelligible* referent, what Jean Borella characterizes as *semantic*.[13] What ultimately validates astrological science reduces thus to the ontological fact that the reality of all that is *sensible*—be it macro- or microcosmic—pertains to the *intelligible* order, where there is no more separation of parts, and where even opposites are subsumed in the unity of wholeness.

So far we have considered the reading of the natal horoscope. There is also, however, an astrology based upon two horoscopes: a natal namely, and one corresponding to a later time, the objective—at least as popularly conceived—being to elicit the "influence of stars and planets" upon the life and prospects of the person specified by the former at the time of the latter. Having previously noted that there *is* actually no such *horizontal* causation, it is evident that this second astrological scenario is likewise to be understood in terms of "wholeness"—which is, after all, what astrology is about. Yet regardless of the cause, the effect presents itself indeed as an "influence" of some kind. The point, however, is that it is *we* who respond to that influence: our human freedom—our own *wholeness* thus—is in no wise abrogated or compromised. On the contrary: it is presented with an opportunity, as it were, to assert itself—in the fullness of its sovereignty no less—by way of its response. We need to understand that, in the sphere of astrology, we are dealing with "whole acting upon whole," as opposed to "parts acting upon parts." And this means, of course, that the "action" too is of a fundamentally different kind: "vertical" as distinguished from "horizontal."

13. We have dealt with this question in chapter 3.

It is to be noted that this "indeterminacy"—or better said, "*freedom*"—does not deprive the astrological prediction of "times and seasons" of its significance or utility: for the better we understand what it is that confronts us, the more enlightened and effective can be our response. Without such an understanding of what "predictive" astrology is actually about, on the other hand, the practice of "peering into horoscopes" may indeed prove harmful: as with every science, one needs to know what one is doing. And therein lies the quandary of astrology ever since the Enlightenment: it happens that the *Zeitgeist* itself impedes us from understanding what that science is actually about. It may thus be "a happy fault" that most people in our day have bought into the "exploded superstition" narrative: it saves them from dabbling in matters they are bound to misconceive.

Meanwhile, of course, astrology remains what it is. I am reminded of an incident in the life of Niels Bohr: he was taking a friend to a mountain cabin he owned, when the visitor noticed a horseshoe mounted upright above the entrance. "Surely, Professor Bohr, you don't believe this can have an effect?" asked the visitor. "Of course I don't," replied the great physicist; "but they say it works, whether you believe in it or not." Well, so does astrology. And despite the prevailing incomprehension on the part of scientists, there are notable exceptions even in that quarter; let me cite at least one example: that of Johannes Kepler—the founder, if you will, of modern astronomy, and a confirmed heliocentrist no less. Referring to astrology he writes: "More than twenty years of practice have convinced my rebellious spirit of its validity."[14]

～

In conclusion I wish to point out that astrology and modern physics are actually related: they prove namely to be *polar opposites*. For whereas physics is the science at which one arrives by reducing the corporeal world to its so-called "atomic" or "particulate" constituents, astrology operates by an implicit reference to maximal wholeness:

14. Quoted by Louis Saint-Martin in *Sagesse de L'Astrologie Traditionnelle* (L'Harmattan, 2018), p. 15.

one completes the corporeal world, so to speak, in the tripartite wholeness of which it forms the lowest tier. This "ontological reversal," moreover, entails a corresponding etiological shift: from horizontal to vertical causality namely, as we have noted. Physics and astrology constitute thus two opposite extremes: *no vertical causality* in the one, *no horizontal causality* in the other. Or to put it in ontological terms: on the one hand a science based upon the "sub-existential" quantum world, and on the other a science based on the "supra-existential" realm Platonists refer to as the *intelligible*.

We should not fail to point out, moreover, that there is yet a third categorical opposition between the two sciences: for whereas both are in a sense "mathematical," the respective kinds of mathematics constitute, once again, diametrical opposites—a point touched upon earlier. And this proves to be crucially significant. We have forgotten that mathematics *per se* does not reduce to its post-Enlightenment genres: that there exist forms of mathematics tracing back at least to the Pythagorean tradition, in which arithmetic and geometry are irreducibly separated, and "numbers" consist simply of integers and their ratios.[15] What confronts us in this long-forgotten pre-Enlightenment mathematics is a science in which "number" does not reduce to sheer quantity, but retains a *qualitative*—one might actually say, an *ontological*—significance. And as we have noted, that mathematics was in fact cultivated vigorously in the philosophical schools—particularly the Platonist—and is intimately connected with the arts, especially with music and architecture. It is the mathematics that speaks to us in the fugues of Johann Sebastian Bach and in the great cathedrals of Europe, the kind that inspired Jean Mignot to say: "*ars sine scientia*

---

15. It may be of interest to recall that "the quantification of geometry"—and hence the validation of "irrational" numbers—was accomplished by none other than René Descartes through the introduction of what to this day is termed a "Cartesian" coordinate system. To appreciate the height and ontological scope of the ancient mathematical sciences it is thus imperative to relinquish the Cartesian opus in its entirety.

nihil."[16] My point is that astrology too rests upon a mathematics linked to the qualitative realm, as exemplified so strikingly in musical harmony, where as we have noted, a halftone can take you "from the world of the Sun to that of the Moon"—and who knows even what this means, until he has "heard" it in the depth of his soul.

If modern physics and astrology—the science of infinitesimal parts and the science of maximal wholeness—prove thus to be in every respect "polar opposites," is it any wonder that in a physics-dominated civilization, astrology should not be held in high esteem? The fact, moreover, that this "science of wholeness" is "incurably Platonist" serves in itself to alienate the physicist, moored as he tends to be in a post-Galilean ontology which constitutes its very negation. It needs to be understood that the current relegation of astrology to the status of a pseudoscience, so far from being based on scientific evidence, rests squarely on pseudo-philosophical grounds: upon the very premises, namely, that define the *Zeitgeist* of our time. Admittedly philosophy as an academic discipline carries little weight these days, and hardly touches upon metaphysical issues anymore; yet in its capacity to close our eyes to the existence of higher spheres, its prowess remains undiminished since the epochal opus of Locke, Hume, and Kant.

Finally let it be noted that not only does astrology have a rational basis—of the loftiest kind, as we have come to see—but it offers a service to humanity no other science is able to provide: a *bona fide* knowledge, namely, of *who we are*; for that is, after all, what—in its own way—the natal horoscope entails. In this regard astrology fulfills a function which has nowadays been taken over by various contemporary schools of psychology, in a manner that is not only upside-down but dangerous in the extreme, as I have argued elsewhere.[17]

---

16. "Art without science is nothing." Jean Mignot was one of the architects of the Milan Cathedral; and there can be no doubt that the heart of his *scientia* was a basically Pythagorean mathematics. If we think of architecture as a kind of "music set in stone," what confronts us, once again, are *harmonic ratios.*

17. See *Cosmos and Transcendence* (Angelico Press, 2008), chapters 5 and 6, dealing with Freudian and Jungian psychology, respectively.

Astrology, finally, can open our eyes to the undiminished grandeur of the universe: the fact, first of all, that the latter is actually a *cosmos* as distinguished from a helter-skelter of particles "*moving endlessly, meaninglessly*" as Whitehead laments. Contrary to what our contemporary sciences lead us to believe, astrology teaches that things exist—not in truth "from the bottom up"—but just the other way round: "*from the top down*" namely. It is not actually "the parts that make the whole," but it is in truth *the whole that makes the parts*. As for myself, moreover, I harbor no doubt that the *Weltanschauung* upon which astrology is based—and which in turn it expresses—is not only true, but normative for mankind at large: that in fact it is consonant with the veritable *sophia* that inspires and enables not only art but human culture in all its higher modes, and ultimately permits us, *Deo volente*, to glimpse—"*as through a glass, darkly*"—the unspeakably glorious calling and last end of human birth.

8

EVOLUTIONIST SCIENTISM: DARWINIST, THEISTIC, AND EINSTEINIAN

IT MAY BE WELL to begin by reminding ourselves that *science* is closely allied with a variety of credenda which are not "scientific" at all, and might consequently be dubbed "scientistic" by virtue of that association. But whereas the adjective may be a neologism, the corresponding substantive "scientism" has by now elicited a Wikipedia article[1] recounting the different shades of meaning assigned to it by contemporary scholars. The list commences with Gregory R. Peterson, who distinguishes two kinds: scientism as a so-called "totalizing view of science," or alternatively as what he terms "a border-crossing violation," which is to say: an illegitimate intrusion of one scientific domain into another. And there is F. A. Hayek, who alludes to "slavish imitation of the method and language of Science"—doubtless a point well taken—and Karl Popper, the eminent philosopher of science, who perceives scientism as "the aping of what is widely mistaken for the method of science." Yet noteworthy and apt as all these attestations to the cultural hegemony of Science may be, they pale into insignificance in comparison to an effect not mentioned at all: the impact of Science, namely, that strikes at the heart of "the human" in us all *by negating its very possibility*. And that is what has been left out of account by the pundits, what in fact "is never mentioned in polite society," as

1. https://en.wikipedia.org/wiki/Scientism

Ananda Coomaraswamy might say: an assumption which proves to be well-nigh invisible by virtue of the fact that it defines the very *Zeitgeist* of the present age. It is therefore *this* conception of "scientism," precisely, that needs to be unmasked above all—on pain of falling victim to its spell.

Thus—opposed though I am to "border-crossing violations," "slavish imitation," as well as "aping" of every kind—my use of the term "scientism" differs radically from the Wikipedia varieties in that I accuse the scientific establishment of an imposture, be it conscious or not: the dissemination, namely, on the basis of what are falsely claimed to be *scientific* grounds, of a *Weltanschauung* which is not only injurious, but ultimately *lethal* to our humanity. That worldview, I charge—currently inflicted upon the young in every science classroom of our land—proves in point of fact to be not only ungrounded and untenable, but insidious in the extreme: for it attacks not only the perennial wisdom of mankind, but our very *sanity*, somewhat as acid corrodes metal.

What, then, is that monstrous doctrine: what precisely does it affirm? I strove to define it on page one of my first book:[2] it rests upon the notion that the universe reduces to purely *quantitative* entities, subject to strict mathematical laws. It hardly matters whether we think of these ultimate components in Newtonian or quantum-mechanical terms: what confronts us either way might be described as "an engine which consists of raw masses wandering to no purpose in an undiscoverable time and space, and is in general devoid of any qualities that might spell satisfaction for the major interests of human nature, save solely the central aim of the mathematical physicist," as Edwin A. Burtt observes.[3] What however renders that opprobrious philosophy virtually irresistible to the contemporary public at large is the fact that it is promulgated by the scientific establishment: the purveyors of "*signs and wonders*" that could "*deceive*" just about everyone.[4]

---

2. *Cosmos and Transcendence* (Angelico Press, 2008); first published in 1984.

3. *The Metaphysical Foundations of Modern Physical Science* (Humanities Press, 1951), p. 299.

4. Mark 13:22

Getting back to the question of "sanity," it is to be noted that the scientistic reduction of the world to *quantity* plunges the thus-deceived into a state of chronic schizophrenia; for obviously everyone takes the grass to be green right up to the moment when he recalls the scientistic credo, at which point he negates what an instant before he took to be self-evident: in a trice the grass has no color anymore! One cannot but wonder on which side of the divide a father, for instance, stands when he hugs his child! What saves us, apparently, from outright madness is that, at most, we only half-believe what present-day science has to tell—and so we end up, in a way, believing nothing at all: the descent into what sociologists term "post-modernity" was predictable from the start.

What I wish, however, to consider are not the adverse consequences of the scientistic worldview—the descent, ultimately, into postmodernist nihilism—but first, *whether that Weltanschauung is justified on scientific grounds*, and above all, *whether it is in fact true*. As to the former, no great wisdom is called for to recognize that the worldview in question is by no means authorized on empirical grounds: that it is in fact *a philosophical postulate masquerading in scientific garb*—which is of course precisely what our usage of the adjective "*scientistic*" is meant to connote. Which brings us to the second issue: whether that scientistic claim is in fact true. And this is a question that can only be resolved on authentic metaphysical ground, which in my view is to be found nowhere but in the great sapiential traditions of mankind: the Platonist, for instance, or the Semitic culminating in the Christian. Yet on that perennial basis the answer is abundantly clear: the ontology of "raw masses wandering to no purpose" proves to be *incurably fallacious*.

∾

Let me attempt to convey what an unspeakably joyful revelation it is that in place of "raw masses wandering to no purpose" we find ourselves actually in a world replete with sound and color and a host of other sensible qualities: the very world artists have painted and poets have sung, which turns out *not* to be a fantasy—a mere *res cogitans*—but an authentic ambience, inhabited by beings we can perceive, communicate with, and *love*. One discerns the stupendous impoverishment and dehumanizing impact inflicted upon present-day civilization through the imposition of the scientistic *Weltanschauung*, not to speak of the concomitant insanity: a heavy price to pay for a dogma that isn't *true*!

However, in addition to this pivotal credo,[5] there are yet other scientistic dogmas, likewise imposed upon us in peremptory fashion, which need also to be unmasked: and undoubtedly the very next in order of importance is the virtually hallowed dogma of *evolution*. If the adjective "sacred" has any meaning at all in the context of professed science, that would undoubtably be the most *sacred* allegedly "scientific" dogma ever enunciated! It needs however to be realized, first of all, that despite the brave talk and unending encomiums lavished upon this darling, the Darwinist tenet is actually bereft of factual support. In the early days—before the middle of the last century, one might say—the Darwinist claim, though improbable in the extreme, was yet scientifically conceivable: one simply did not know enough about biology to rule it out. After the mid-century discovery of DNA, on the other hand, followed by the publication, in 1998, of William Dembski's mathematical theorem concerning the production of CSI or complex specified information, such is however no longer the case: for whereas the first discovery reveals massive amounts of CSI in the nuclei of living cells, the latter proves that this cannot be produced by natural

---

5. The ontological dichotomy, to be precise, which underlies and supports the scientistic worldview in all its facets: i.e., the Cartesian division of reality into *res extensae* or "extended entities" and *res cogitantes* or "things of the mind," what Alfred North Whitehead refers to as "bifurcation." Highly respected in scientific circles as co-author, with Bertrand Russell, of *Principia Mathematica*—the definitive treatise on the foundations of mathematics—Whitehead lectured the scientific community for decades on the fallaciousness of their "bifurcationist" worldview, yet evidently to no avail: the ontological fallacy appears to have deep roots, tracing back to another *Principia* in fact: the one published in 1687 by Sir Isaac Newton, from which modern physics is derived.

means.[6] From that point onward, Darwinian evolution can no longer be viewed even as a viable scientific hypothesis, what to speak of a "scientific" fact.

Not, to be sure, that the theory was consequently abandoned wholesale or replaced! As Thomas Kuhn makes abundantly clear, science as a social phenomenon does not operate that way: a theory of paradigmatic rank is not normally abandoned when it has failed. "Once it has achieved the status of a paradigm," he writes, "it is declared invalid only if an alternate candidate is available to take its place."[7] And in the case of Darwinism, to be sure, none is in sight.

∼

Almost none, that is; for it happens that a turn of events no one expected has come to pass: the emergence, namely, of a "theistic" genre of Darwinist evolution, which appeared—as if *ex nihilo*—in the early decades of the twentieth century. Let us briefly recall that unlikely chain of events.

Darwin himself, to be sure, was not "theistically" inclined, and his theory, as presented in his *magnum opus* of 1859, was scientifically orthodox "by the book." Having observed morphological discrepancies between mainland specimens of certain species and their kindred on the Galapagos Islands, and having drawn inductive conclusions based upon these facts, he went so far as to submit his theory to the Popperian condition of falsification. Thus, in the *Origin of Species*, he asserts that if the apparent paucity of "fossiliferous deposits" in the Precambrian strata should prove to be

---

6. By "natural causes" one understands a deterministic, random, or "stochastic" process, the last being a combination of the first two. The Darwinist scenario consisting of random mutations plus "natural selection" is a case in point. It cannot therefore produce the "tons" of CSI contained in the nucleus of every cell. Despite persistent denial on the part of the scientistic "faithful," the matter is actually as simple as that.

7. Kuhn's thesis is that a fundamental paradigm is never discarded until a replacement has been found. See *The Structure of Scientific Revolutions* (University of Chicago Press, 1996), p. 77.

factual, this could "truly be urged as a valid argument against the views here entertained."[8] Yet within a century, Darwin's worst fear was in fact realized: the "intermediary forms" which the theory evidently demands had by then proved not to exist. "Most species exhibit no directional change during their tenure on earth," writes one of the leading authorities: "They appear in the fossil record looking pretty much the same as when they disappear."[9] The status of Darwin's theory as a viable hypothesis had thus been disproved; as Phillip Johnson notes: "Darwinism apparently passed the fossil test, but only because it was not allowed to fail." In a word, one had arrived at the Kuhnian scenario, which is to say that Darwinism had morphed from a *scientific* hypothesis into what I term a *scientistic* dogma, accepted henceforth with well-nigh "religious" homage by the scientific community as well as the public at large.

From this point onwards, Darwinist evolution was no longer an hypothesis to be verified, but had become a dogma to be justified: science had turned into politics one might say. Soon enough the theory was embraced in various quarters as something to be disseminated and imposed as widely as possible, and ere long was taught in schools and universities throughout the land as an established fact. It appears the transition from scientific hypothesis to scientistic dogma took place long before the Darwinist hypothesis had in fact proved to be untenable, and has not been visibly affected by that recognition. The transition from an hypothesis to a dogma is moreover attested by the *modus operandi* used routinely to neutralize adverse findings: whenever, namely, facts came to light which appeared to falsify the theory, the experts set about—with astounding ingenuity—to "explain away" the falsifying facts by means of this or that *ad hoc* postulate. To cite an extreme yet by no means untypical example: to account for the virtual nonexistence of "intermediary forms" in the fossil record, someone actually proposed—with a straight face before a Congress on Philosophy

8. *On the Origin of Species* (Chicago: Britannica, 1952), p. 164.

9. Stephen Gould quoted in Phillip Johnson, *Darwin on Trial* (InterVarsity Press, 1993), p. 50.

of Science in 1949[10]—what he termed the "automatic suppression of origins." We have here what might be termed the poster-child of that emerging strategy, designed to protect a scientific theory against falsification by means of an additional hypothesis literally "pulled out of thin air."

Meanwhile something stranger yet was in the offing: a second metamorphosis—which even writers of fiction might never have imagined—was about to transpire: ere long, namely, the Darwinist doctrine had assumed a *theistic* form as well. One way or another, it seems, the concept of "evolution" would establish itself as the presiding paradigm! And as one knows, it was the charismatic Jesuit paleontologist Pierre Teilhard de Chardin who put "theistic evolution" on the map through the well-nigh hypnotic impact of his writings. Ignoring the customary premises of theology and science alike—not to speak of the Index in force at the time—his long-proscribed books spread like wildfire in Catholic circles around the time of the Second Vatican Council, the outcome of which they influenced decisively. Perhaps the most striking feature of the Teilhardian opus is the unprecedented use of intoxicating metaphors which appear to have mesmerized millions—up to the most highly educated strata of Western society. Peter Medawar speaks for the wary minority when he avers that "Teilhard habitually and systematically cheats with words": for instance when he employs scientific terms in contexts to which they absolutely do not apply: an "offence against the common decencies of scientific writing" the Nobel Laureate calls it.[11] Little however did Sir Peter realize that the legacy of Teilhard de Chardin would remain a part of our culture after his "cheating words" had long been forgotten. What the charismatic Jesuit brings to the scene is a concept of "evolution" which, so far from being atheistic, brings God himself into the evolutionist scenario. And whereas this "revelation" was generally received as more or less innocuous by mainstream Darwinists—presumably because it struck them as utterly absurd—to

10. The paleontologist in question, incidentally, was Teilhard de Chardin, regarding whom we shall have more to say.

11. On this entire subject I refer to my monograph, *Theistic Evolution: The Teilhardian Heresy* (Angelico Press, 2012).

the embattled believers it was sheer *manna* sent from above. No wonder a chorus of cheers went up from behind the ramparts!

In due time the doctrine of "theistic evolution" established itself in various segments of the educated world, and in Vatican circles attained in effect the status of a new orthodoxy. The dominant segment of the Catholic intelligentsia, it appears, had by now become persuaded that God "creates" through evolution by supplying the final "push" to effect a *bona fide* Darwinian "origin of species"—and presumably pull off, as an additional miracle, that "automatic suppression of origins" affirmed by the Master himself.

The story does not, however, end with "theistic" Darwinism; for it happens that, in the course of the twentieth century, another kind of "evolution" has made its appearance: an evolution of the universe itself, but based this time—neither on morphological observations nor on pseudo-theological speculation—but on the most sophisticated mathematical physics the world has ever seen: the Einsteinian, namely. Yet a very similar pattern of "crisis begetting *ad hoc* intervention" has come into play: the underlying strategy, in fact, is quite the same. An impasse presents itself: in the Darwinist scenario, it may be a "missing link," where in the astrophysical it is an insufficiency of matter, say, to produce gravitational fields strong enough to account for the formation of stars and galaxies. And in either case the impasse is broken by means of an *ad hoc* postulate: an assumption "picked out of thin air." But whether the added ingredient be an "automatic suppression of origins" or "dark matter," the logic is identical. One might note that, in point of fact, the list of these "discoveries" proves to be by far longer in the case of astrophysics: for in addition to such major innovations as "inflation," "dark matter" and "dark energy," an abundance of less celebrated "entities" of *ad hoc* provenance is to be found in the technical literature. As one insider has observed with unusual perspicacity: "Every time there is a new observation there is a new theory!"[12]

---

12. The astronomer Brent Tully of "supercluster" fame.

Yet despite the prevailing *carte blanche* to "*ad hoc*" one's way out of empirical quandaries, discrepancies have cropped up which prove immune to this strategy. For example: having added "dark energy" to the inventory of astrophysical ingredients, it turns out, as we have noted before, that the corresponding vacuum energy conflicts with the findings of quantum theory by a whopping 120 orders of magnitude! This is by far the most gargantuan "miss" in the history of science—and no *ad hoc* acrobatics nor "god of the gaps" can ameliorate this fact.

Incredible however as it may seem, worse was yet to come. For it happens that the entire edifice of Einsteinian astrophysics rests upon an assumption of homogeneity "in the large" known as the Copernican or Cosmological Principle; and guess what: this tenet too has absolutely failed! It has been disproved, moreover, on the basis of astrophysical observations of the most accurate kind, culminating in tons of high-precision data transmitted from the Planck satellite, launched in 2009. What Planck documented beyond all reasonable doubt was the existence of a global "axis" in the CMB or "cosmic microwave background"[13]—the largest and supposedly oldest identifiable structure in the universe—which flatly negates the aforesaid Principle. The fact is that in an Einsteinian universe there *can be* no such "axis"! And logically speaking, this settles the matter once and for all: one can't "*ad hoc*" against a flat contradiction. A point has been reached where "all the king's horses and all the king's men" simply can't put the Einsteinian theory "back together again." No wonder that at the very first glimmer, that ominous structure in the CMB was dubbed "the axis of evil"! And as a matter of fact, Planck was designed specifically to disprove its existence, and to this end was equipped with the most advanced instruments conceivable, designed to filter out every trace of spurious data from every possible source—but all to no avail: at the end of the day, there it was, that "wicked" axis—*the size of the observable universe* no less—staring the astrophysicists in the face.

---

13. W. Zhao & L. Santos, "The weird side of the Universe: Preferred axis," *International Journal of Modern Physics: Conference Series*, Vol. 45 (2017).

⁓

The question remains how "evolution" as such—freed of Darwinist, Teilhardian, or Einsteinian assumptions—stands from an authentically theological or metaphysical point of view: and in either case it is by no means difficult to discern that, in truth, *it does not "stand" at all.* To begin with the theological: the closest a Christian text ever comes to the idea of an "evolution" is in the context of the so-called *logos spermatikos* or *ratio seminale*, where an "evolution" of sorts is in fact implied.[14] What stands at issue, however, is manifestly an "evolution" in the literal sense of an "unfolding" of something that already exists: of a *logos spermatikos* or *ratio seminale* namely. And that "unfolding" constitutes moreover a *vertical* as distinguished from a *horizontal* process: an ontological *descent* one can say, not "in time," but "into time and space" precisely. Or to look at the question from a somewhat more "exoteric" and overtly Biblical point of view, the plain fact is that God created the plants and animals "*each after its kind*": He did not create a fish, for example, to morph into a frog or a chimpanzee! This is not to say, of course, that the primordial types have all survived to the present day in their original forms: there can be no question that variations and adaptations have taken place, giving rise to a different panorama of living forms. My point, however, is that these alterations pertain to what is properly termed "microevolution," the reality of which no one doubts: it is what Darwin documented so clearly in the *Origin of Species.* And whereas even experts may not agree on just how far these "microevolutionary" variations can reach, there can be not a shred of doubt that they do not stretch from a fish, say, up to a frog, let alone a chimpanzee. These are fantasies, which would doubtless be quickly waived aside by the experts if it were not for the fact that they are more, by far, than a mere conjecture or scientific "hypothesis": the point is that these postulates prove to be ideology-driven, that at bottom we are dealing with something inherently *religious*—or *anti*-religious, as one can likewise say.[15]

14. See above all St. Augustine, *De Genesi ad Litteram.*
15. We shall return to this question in chapter 11.

Shifting now from theology to metaphysics, one finds that the case for "evolution" stands no better, and in a sense, even worse. For if—starting so to speak "at the top"—we recall that the Platonist tradition places "the real" on the plane of the *intelligible*, the tenet of evolution is literally "dead on arrival": for on that plane there can be no "evolution" because *there is no time*—no change or alteration of any kind. Let us shift then from Plato to Aristotle—the first authentic scientist, one can say. It may therefore come as a surprise that the Aristotelian ontology—which proves to be hylomorphic—turns out to be no more "Darwin friendly" than the Platonist: for whereas it obviously permits *change*, it nonetheless excludes evolution categorically. A fish, for example, derives not only its being but its "fishness" from a *form*, a *substantial form* to be precise. To morph into a frog or a chimpanzee, that *form* itself must consequently morph: but it doesn't, *it can't*—for the simple reason that its own essence derives from the intelligible realm where "time is no more." One sees that, in the final count, Aristotle is still a Platonist: reality stems yet from the intelligible world, the aeviternal realm of archetypes in which there is no change nor "*shadow of alteration*"—not the slightest trace, thus, of "evolution."

The conceptual possibility of "evolution" in the contemporary sense commences historically with the Enlightenment,[16] which has in effect turned the world upside-down: instead of the part deriving from the whole, the new philosophy conceives the whole as no more than an assemblage of parts. And on *that* basis, to be sure, evolution becomes instantly conceivable: all one need do in principle to produce a frog or a chimpanzee, say, from a fish, is "jiggle" the parts suitably, and *voilà*! The details do of course get ever more complicated, and this very complexity tends in turn to fascinate and draw the seekers ever deeper into the futile quest. Yet the tiniest dose, even, of Plato or of Aristotle suffices in principle

16. To those who would point to Democritus, for example, I say first of all that his doctrine was regarded during the golden age of Greek philosophy as a heresy, and not until the Enlightenment, some two thousand years later, was it revived. And true enough, it has since then become in effect the ontological basis of contemporary scientism in all its manifestations.

to induce a *metanoia*: bring us back into the real world, consisting not only in a horizontal plane and a "below," but endowed with an "above" as well. A few pages of the *Timaeus*, in fact, should by right suffice to save us from ever setting foot in that "evolutionist" maze from which—as from the Cretan labyrinth—those who enter seem fated never to return.

9

VERTICAL CAUSATION
AND WHOLENESS

VERTICAL CAUSALITY made its appearance in the context of the measurement problem in quantum mechanics, where it could be identified by the fact that it acts "instantaneously."[1] Whereas the previously known modes of causation—subsequently referred to as "horizontal"—operate *in time* by way of a transmission through space, vertical causality operates directly, without the mediation of such a process. That "instantaneity" or lack of transmission came thus to be taken as the defining characteristic of vertical causality. But whereas this criterion may serve to identify VC, it does not tell us whence it acts and what it effects. It is time, now, to broach these deeper questions: time to delve into the metaphysics of VC, in the hope this may shed light as well upon questions of scientific significance. I begin, then, with the definitive claim that vertical causation is nothing more—and nothing less—than *the causation effected by wholeness.*

Which brings us to the crucial recognition that *the very existence of VC—i.e., the causation effected by wholeness*—entails that *physics is not the "all inclusive" science it is generally assumed to be.* The very existence of vertical causation rules out such a thing as that "theory of everything" particle physicists have been laboring for almost a century to formulate. The appearance of VC—at the exact boundary between what I term the *physical* and the *corporeal* domains—puts an end to that expectation. To which I would add that this boundary is likewise the only juncture at which VC becomes in

---

1. See *Physics and Vertical Causation* (Angelico Press, 2019), ch. 3.

a sense *measurable*: for only in a transition between *two distinct ontological domains* can "instantaneity" be empirically verified.[2] It is thus at the very instant of that *ontological* transition that an undeniable footprint of VC can be discerned.

The rediscovery of what I term the *corporeal world* proves to be tantamount to the rediscovery of *wholeness*. But whereas the very idea of wholeness comes as a shock to the physicist—given that physics is in fact based upon its denial—its re-emergence opens the door to the rediscovery of a perennial insight blocked since the Enlightenment. For it has ever been known to the wise that, in the absence of *wholeness*, there is no *being* as well: "*ens et unum convertuntur*" declares a Scholastic dictum—which signifies in effect that *being and wholeness are one and the same.*

~

Given then that we live in a world not made of quantum stuff, but composed rather of wholes, let us ponder the notion of "vertical causation" simply as an *act of wholeness*. The moment we do so, however, the realization dawns that a wholeness subject to the conditions of space and time will not suffice: for whether we are presently able to grasp the point or not, the ontological fact is that a corporeal wholeness—a wholeness subject to the bounds of space and time—is perforce *secondary* or derived. And let me add that this fact should not altogether surprise us, given that vertical causation acts *instantaneously*: for does this not suggest that the wholeness from which it acts is not subject to the condition of time?

The fact is that primary wholeness is not to be found on the corporeal plane, but pertains rather to that Center to which Dante alludes as "*il punto dello stelo a cui la prima rota va ditorno*":[3] that mysterious "Point" around which "the primordial wheel" is said to "revolve." To speak in philosophical terms, this signifies that it is

2. When it comes to ordinary measurement, the distinction between "instantaneous" and "exceedingly fast" cannot be empirically ascertained. On this question see *Physics and Vertical Causation*, op. cit., pp. 26-9.

3. *Paradiso* xiii, 10.

needful to recover the ontological conception of the *tripartite cosmos*, which is centered precisely upon that *punto dello stelo* where space and time are transcended. In the final count, metaphysics is indeed a matter of viewing the integral cosmos from the vantage of that Pivot around which time itself is said to circulate—in keeping with Plato's enigmatic reference to time as "*the moving image of eternity.*"

This is not of course "philosophy" as innocuously understood in our day, but rather as comprehended by the wise in premodern times, which is ultimately a matter—not of conceptual speculation or clever reflections concerning language—but of a higher kind of *seeing*. And so long as we are obliged to see "*as through a glass, darkly,*"[4] it behooves us to employ an appropriate icon for our "glass"— which brings us back to what I have termed "the cosmic icon," upon which, in a preceding publication,[5] I based my discussion of vertical causality. The integral cosmos comes then to be represented by a circle which breaks into three domains: the center, the interior, and the circumference. Now the key which renders this schema enlightening ontologically relates to the cosmic bounds of *time* and *space*, which need to be assigned to the two outer regions in the correct order, the point being that *time has precedence over space.*

What proves crucial for metaphysics at large is the iconic fact that the center of the cosmic icon—or better said, the ontological domain which it represents—is subject to neither bound, and corresponds thus to Dante's *punto dello stelo*. And to speak in inherently Platonist terms: that is the Center where, ultimately, *all wholeness resides*. Its opposite—the circumference namely, representing the corporeal world—rests, as it were, upon that Center: *for it is from thence that such wholeness as it contains is derived.* Finally, let us note that these two extremes—the aeviternal[6] Center and the corporeal Circumference—are joined by an intermediary ontological domain, which is subject to *time* but not to *space.*

4. 1 Cor. 13:12

5. *Physics and Vertical Causation*, op. cit., ch. 8.

6. The best definition I know comes from St. Thomas Aquinas: "Aeviternity itself has neither 'before' nor 'after,' which can however be annexed to it." *Summa Theologiae* I, Q. 10, Art. 5.

Now, as I have noted elsewhere, this intermediary domain—integral to the traditional schools of metaphysics, and in a way known also in occult circles—has been roundly forgotten in the contemporary world.[7] No *bona fide* metaphysical comprehension of the corporeal realm is possible, however, without a recognition of that intermediary world: for it happens that corporeal entities do not stand alone, but derive such wholeness as they embody from the aeviternal realm *by way of the intermediary*. The fundamental fact—upon which all authentic metaphysics rests—is that *cosmic reality is inherently tripartite*, and that *all authentic wholeness*, be it on the corporeal or on the intermediary plane, *derives "from above": from that Center beyond the pale of both space and time.* And let us note, as a corollary, that *vertical causation*—which is the causation exercised by wholeness—*emanates perforce from the aeviternal plane*, which is after all the reason why *it acts*, not "in time," but *instantaneously*.

∾

Primary wholeness, then, resides "above," in that supreme ontological realm to which the Platonist adjectives *aeviternal, archetypal,* and *intelligible* properly apply. Yet "esoteric" and "empirically irrelevant" as that ontological conception may appear to the contemporary mind, it happens to be essential to the resolution of our scientific quandary. For as we have by now come to see, what limits the scope of physical science—and ultimately relegates the so-called "physical universe" to the status of a sub-existential domain—is the fact that physics has no conception and no grasp of *wholeness*, and in fact operates by systematically breaking down all corporeal wholeness in quest of Democritean atoms or quantum particles, which prove in the end not to exist.

7. Still remembered in the Orthodox Church as the so-called "aerial world," considered to be the habitat of demons, it was termed the "astral plane" in 19th-century occultism. I might mention that, in our day, Malachi Martin became rather well-acquainted with that domain in his capacity as an exorcist, and referred to it habitually as the "middle plateau."

It may be noted at this juncture that a *Weltanschauung* based upon physics cannot but be evolutionist to the core: for once authentic wholeness has become *de facto* inconceivable, it needs to be replaced by a pseudo-wholeness which reduces to the sum of its parts. And needless to say, the only way such a pseudo-wholeness can conceivably emerge is by an aggregation of its constituents. Moreover, inasmuch as such a pseudo-wholeness exemplifies no model or archetype, its formation cannot be in any sense "directed," and reduces therefore basically to a random process. In a word, *Darwinian evolution is essentially the one and only means by which plant and animal forms could conceivably originate in a world answering to the conceptions of physics.* And this in itself accounts for the fact that the Darwinist hypothesis has been doggedly retained in the face of persistent empirical failure, and even after it has been mathematically disproved, in 1998, by William Dembski's theorem to the effect that *horizontal causality cannot produce "complex specified information" or CSI.* Let me emphasize that inasmuch as the nucleus of every living cell—of even the most primitive organism—literally "teems with" CSI, we now know for certain that *it takes vertical causation to produce a living organism.* Moreover, this in turn entails, and again in light of our reflections, that *a living organism constitutes*—not a mere aggregate or Darwinist pseudo-whole—but *a bona fide whole.* And again, in light of our reflections, this further entails that such an organism can, in turn, *act as an agent of VC* in its own right. Finally, it has become clear by now that the attempt to understand living organisms on the basis of physics cannot but fail in the end.

Getting back to wholeness as such, the ontological fact is that what I term the *corporeal* world is indeed made up of entities such as "red apples," which prove not to be mere aggregates, but *wholes.* The fact that they are *divisible,* in other words, does not entail that they are in truth *divided*—that they are *aggregates* rather than *wholes.* What ontologically distinguishes the *corporeal* from the *physical* realm is in truth the fact that the former is comprised of wholes, whereas *there is no wholeness in the quantum world.* It is crucial to note, moreover, that this explains why *vertical* causation enters the picture precisely in the act of measurement: for inasmuch as measurement constitutes a transition from the physical to the corporeal plane, it entails a transition to wholeness, which horizontal causality cannot provide. And that is why vertical causation enters—and *must* enter—into play at the instant of that ontological transition.

The pieces of the puzzle are beginning to fall into place, which is to say that the "big picture" is beginning to emerge. It turns out that the corporeal world is composed, not of quantum particles, but indeed of *wholes*—and thus, in a sense, of the very opposite. The question now presents itself: what are these wholes, and whence do they arise? And strange as it may seem, what is called for at this crucial juncture is none other than the Platonist recognition that *"wholeness resides ultimately in the aeviternal realm."* This connects, first of all, with what, for us, has been from the start the defining characteristic of VC: the well-nigh incomprehensible fact, namely, that *vertical causation acts instantaneously.* For this, in itself, is indicative of the fact that *VC does not originate in this, our time-constrained world.* No physicist, after all—however ingenious—is capable of producing an "instantaneous" effect, nor could that "instantaneity" be measured or verified by empirical means. The fact—the *momentous* fact, to be sure—is that both *vertical causation and wholeness originate from that punto dello stelo* no less, where the dispersion of time does not reach.

The definitive characteristic of Platonist metaphysics resides in the recognition that all temporal being or wholeness stems ultimately from an *aeviternal* archetype. It appears to be in fact the characteristic of all "higher" wisdom schools not only to recognize an "eternity," but in a way to subordinate the *temporal* to the *eternal,* the *transient* to the *immutable.* What I wish now to point out is that *this very subordination of the temporal to the eternal is in fact the key to the enigma of vertical causation and wholeness.* It first of all explains, as we have noted, why VC acts "instantaneously": it does so because it does not originate "in time." It acts thus in what the Scholastics termed the *nunc stans,* the "now that stands," which proves ultimately to be the only "now" there is: for as St. Thomas Aquinas apprises us with the utmost brevity, *time consists of before and after*.[8] The fact is that the *creation* of the temporal world, whether it be attributed to God or to a Demiurge, takes place—not "in time"—but precisely in that *nunc stans*; as Meister Eckhart says, "God makes the world and all things in this present 'now'."[9]

The action of VC is first of all Demiurgic in that it gives rise to the temporal order, consisting of entities based upon aeviternal prototypes. Whatever *being* we encounter on the corporeal or the intermediary planes constitutes thus a *temporal* manifestation of an *aeviternal* whole. One sees that *the corporeal and intermediary strata of the integral cosmos are brought into existence precisely by way of vertical causation*—which moreover should surprise no one, seeing that *horizontal* causality presupposes the bounds of time and space. Inasmuch, therefore, as horizontal causality operates within a spatio-temporal continuum, it is *an effect of vertical causation*. And this explains why VC has power to override horizontal causality, as in the act of measurement, when the Schrödinger wave equation is "re-initialized" following the so-called "collapse" of the wave function.

The question remains what "causes" the quantum realm: could it be VC? In a sense this is of course the case, for in the absence of VC, there would be no temporal strata of existence, and therefore no quantum world as well. What, on the other hand, speaks against that argument is the ontological fact that, strictly speaking,

---

8. Op. cit., Art. 1. The complete passage is as follows: "As we attain to the knowledge of simple things by way of compound things, so must we reach to the knowledge of eternity by means of time, which is nothing but the numbering of movement by 'before' and 'after.' For since succession occurs in every movement, and one part comes after another, the fact that we reckon before and after in movement, makes us apprehend time, which is nothing else but the measure of before and after in movement. Now in a thing bereft of movement, which is always the same, there is no before or after. As therefore the idea of time consists in the numbering of before and after in movement; so likewise in the apprehension of the uniformity of what is outside of movement, consists the idea of eternity." It was Aristotle who defined time as "the numbering of motion with respect to before and after."

9. *Meister Eckhart*. Trans. C. de B. Evans (Watkins, 1924), vol. I, p. 209.

the quantum world does not exist: for it is made up, after all, of mere *potentiae*, which do not attain being until they are *actualized*. And that, as we know, is precisely the point at which VC enters into play: for this actualization of quantum *potentiae* occurs at the very instant of measurement *as an act of VC*.

~

It appears that vertical causation is the primary causality which brings the temporal orders—inclusive of their sub-existential modes—into manifestation. In this capacity, VC turns out to be none other than what, in Platonist parlance, is termed *formal causation*.[10] Based upon the preceding considerations, moreover, the adjectives *primary*, *creative*, or even *Demiurgic* could likewise be applied to identify this mode of causation. It is to be noted, however, that *vertical causation*, as we have conceived of it, is by no means restricted to *formal* causation in this primary or "creative" sense. As a matter of fact, inasmuch as *horizontal* causality presupposes both time and space—and consequently is operative only within the corporeal, and subcorporeal, domains—it follows that *VC constitutes likewise in truth the primary efficient causality*.

What the primary or creative VC brings into being—to speak now in basically Aristotelian terms—are "substances" determined by a *substantial form*. It needs however to be realized that—so far from acting like marionettes—every such being is endowed with a certain capacity to act upon other beings by a vertical causation of its own, which derives from its substantial form. This *secondary* mode of VC can therefore be termed *substantial*. And let me emphasize: "substantial" or secondary as it may be in comparison to the primary or creative kind, that causality is yet authentically *vertical* inasmuch as it operates "instantaneously," and thus not "in time." How, then, is this possible, given that this mode of VC derives supposedly from a substantial form, which after all

10. On this issue I refer to a doctoral dissertation by Andrew R. Hill, submitted at the Catholic University of America in 2016, under the title "Forms as Active Causes in Plato's *Phaedo* and *Timaeus*."

pertains to the intermediary or "time-bound" realm? What needs to be grasped is that a substantial form, though subject to the flux of time, does not exist in and by itself, but derives its essence and being "from above": that is to say, from its aeviternal prototype. The point is that the substantial form remains ontologically connected to its aeviternal prototype, which is to say that the two constitute a *whole*: an "organismal" whole, one might say. What Plato affirms regarding the soul—i.e., that *"it is partly in eternity and partly in time"*—applies thus to substantial forms at large. So far from being separated or cut off from their supra-temporal prototype, they are in reality a manifestation thereof; a *temporal* manifestation, to be precise.

We began this inquiry with the stipulation that *vertical causation constitutes the causality of wholeness.* It now appears in addition that *wholeness as such is characterized by the capacity to act by way of vertical causation.* Let this much, then, suffice as an introduction to the Platonist view of the integral cosmos.

10

THE MYSTERY OF
VISUAL PERCEPTION

Lord Kelvin made a crucially important point when he declared physics to be "the science of measurement": it is this basis in *measurement* that proves definitive not only of physics and the physical sciences at large, but of the resultant *Weltanschauung*. Obviously measurement *quantifies*: in terms of the standard cgs system, it reduces things in principle to "centimeters, grams, and seconds." The fact, however, is not that all is thus reducible, but that whatsoever happens not to be is "filtered out" by the *modus operandi* of the scientific process and eventually demoted to the status of a *res cogitans*, a mere "thing of the mind." And it is this scientistic denial of the non-measurable that in a way defines not only our science but our culture at large.

It can hardly be doubted that the physical sciences have proved successful beyond the wildest dreams of the founders themselves: even Francis Bacon, I surmise, would be amazed to see what his celebrated "machine for the mind" has been able to accomplish! On the other hand we generally fail to grasp what a physics-based science does *not* accomplish, what in fact its very outlook excludes. A physics-based science, first of all, cannot disabuse us of the belief that the world reduces to *quantities*: to the things, namely, brought into play by that Baconian "machine." What is lost in the bargain are above all the sensible qualities: the five kinds which render the corporeal world perceptible. These sensibly apprehensible constituents are now said to be "subjective," not because some *experimentum crucis* has shown them to be, but simply because a science based upon measurement does not register such a thing as *red* or *green*. Color *per se* has thus been identified as a wavelength, notwithstanding the self-evident fact that what we perceive is not a wavelength at all, but *red* or *green*.

Granting that the world does encompass qualities—and consequently does not reduce to the conceptions of the physicist—the question arises whether there may be other kinds of natural science, legitimate and useful in their own right, which do *not* subjectivize the sensible qualities. But whereas that possibility has in no wise been disqualified, it has evidently been ruled out of court by the presiding authorities. To which I would add that these sciences are founded upon principles one can neither comprehend nor evaluate on the basis of our contemporary *Weltanschauung*, which is why they are nowadays classified as "pre-scientific superstitions." And I venture to suggest that what in these premodern sciences replaces *measurement* is *perception*, which I presume was primarily *visual*.

It needs of course to be understood that this was, most assuredly, no ordinary perception, failing which every normal man, woman, or child would have been a scientist! What stands at issue, evidently, are modes and degrees of perception in excess of our current "normal"—which brings us to a crucial point: the fact is that the requisite degrees of perception generally demand an authentic discipleship, a *bona fide initiation*. And this is something we find difficult, nowadays, to grasp, first of all because *there is no such thing* in our contemporary civilization. Yet it did in truth exist once upon a time, which is moreover the reason why the sciences in question are said to be *traditional*. In a word: what renders a traditional science operative is a *transmission*, a "passing on" of something from *master* to *disciple*.

But such a conception, needless to say, makes no sense whatever in an Einsteinian or a Darwinist universe. It does, on the other hand, make eminent sense if the cosmos at large, as well as the human microcosm, prove to be ontologically tripartite, as we contend they are. It is then not only conceivable but altogether to be expected that a "preternatural" knowing involving the intermediary—or even the aeviternal—plane will have an efficacy "not dreamt of in your philosophy."

Happily, we are in truth greater—by so many ontological "orders of magnitude"—than an Einsteinian or a Darwinist universe would permit us to be: nothing, indeed, has been more underestimated and demeaned in our contemporary civilization than *man* and *woman* as such. Yet the fact remains that the integral human person bears within himself a potential so great that his present intellectual formation renders him incapable of catching so much as a glimpse thereof. Our pseudoscientific schools of psychology, meanwhile, have contributed mightily to what might be termed the "trivialization" of mankind: given that physics has reduced the cosmos to a mechanism, it has *ipso facto* turned the inner world into a subjective realm of libido and illusion. That "subjective world" has thus become fair game to all kinds of geniuses, from Freudians intent upon turning it into a veritable sewer, to Jungians aspiring to create a new "deep-psychological" religion.[1]

Getting back to *visual perception* as the *modus operandi* of the traditional sciences, it is to be noted that not only is man ontologically tripartite like the integral cosmos, but there exists a kinship between the two so profound that the *anthropos* has been termed a "microcosm." And I would note in passing that there is more to the Delphic injunction *"Know thyself"* than could be imagined from our contemporary point of vantage: when it comes to a simian who has but recently learned to walk on his hind legs, the Delphic injunction, admittedly, doesn't sound too enticing. On the other hand, given that man constitutes a tripartite microcosm no less, the picture has changed: it then becomes conceivable that perception can take us in principle to the very heart of cosmic reality for the simple reason that it springs in the final count from that very "heart" itself. It needs thus to be noted that there exist perforce different *kinds* as well as *degrees* of perceptual knowing, and that visual perception in particular can be sharpened to penetrate into hitherto inaccessible vistas like a laser beam. Is it any wonder, then, that the disciples of Pythagoras, for instance, were obliged to observe—in addition to a vow of celibacy—five years of

---

1. I have dealt with this issue at some length in *Cosmos and Transcendence* (Angelico Press, 2008), chapters 5-6.

absolute *silence*: are we able so much as to imagine to what degree such practices can "sharpen the sword" of the human spirit and empower us to *see*! Let me propose, then, that henceforth all reference to "ancient superstitions" be put on hold while we endeavor to ascertain the facts.

~

It is time to realize that *perception* is, finally, a mystery—as are ultimately the entities we perceive. There exist of course officially mandated theories of perception, visual first of all; yet these prove in the end to be untenable.[2] In fact, they *must*, given that a science based upon measurement excludes *a priori* the very elements that render objects perceivable: the *qualities* namely, which—as we have noted repeatedly—have, for the past several centuries, been consigned to the limbo of subjective fabrications. Absent the qualities, however, *there can be no perception*.

What first of all stands at issue thus, in "the problem of perception," is the nature or quiddity of the corporeal realm: the fact that the corporeal domain is endowed with five distinct kinds of qualitative attributes, answering to the senses of sight, hearing, touch, taste, and smell. The point is that these five sensible qualities—so far from being "merely subjective"—prove, on the contrary, to be definitive of corporeal existence as such. What renders a thing *corporeal* is the fact that it can be known through at least one of the five senses: not that it *is* thus perceived, but that in principle it *can* be.

Let us then be open to the time-honored notion that corporeal entities are somehow composed of the "five elements." We need not be put off by the naïve-sounding designations of these so-called elements, traditionally referred to as *earth, water, air, fire,* and "*aether*," the *quinta essentia* or "fifth essence." It hardly needs saying that the "earth, water, air, and fire" in question are not to be taken in the ordinary sense, any more than the *quinta essentia* may be identified with the "ether" once postulated by physicists as the

2. On this point see ch. 5.

medium supporting electromagnetic waves. What proves crucial is a connection between these five purported elements and the five sensible qualities, in the specific order *smell, taste, sight, touch,* and *sound,* namely. It is further to be noted that the aforesaid designations of the "five elements," along with this five-fold correspondence, appears to have been universally accepted in premodern times. An unequivocal imprint of this doctrine is moreover to be encountered from the shores of Egypt and Greece to the Himalayan regions of India and Tibet. It would in fact surprise me if a reputable medicine man among the American Indian tribes, say, did not have at least a smattering of these doctrines as well.

This science of the "five elements" appears moreover to have attained its most explicit and refined formulation in the Hindu *darśana*[3] known as Sānkhya, where these elements appear in two forms: a subtle pertaining to the intermediary domain, known as *tanmātras,* and a gross referred to as *bhūtas,* conceived as being constitutive of the corporeal. It is these five *tanmātras,* then, that constitute the intermediary or "subtle" basis of *corporeal* existence as such. And this should in fact come as no surprise, given that "corporeal" is indeed tantamount to "perceptible," and that it is finally the five *tanmātras* that render entities perceptible according to the five modes. It needs however to be clearly understood that these *tanmātras* pertain incurably to the intermediary realm: no wonder they are invisible to our physics-based means of inquiry!

~

This brings us to a crucial point regarding the nature of perception which needs to be clearly grasped: in whatever mode, *perception is not—and cannot be—consummated on the corporeal plane.* It will be instructive, first if all, to consider the matter from the standpoint

3. This term—derived from the verb *driś,* "to see"—is the closest Sanskrit equivalent of what we term "philosophy," the point of difference being that to the Indian mind there is no single *darśana* that covers the entire ground. Which is to say that one requires, not one, but *six darśanas*—corresponding to the six "directions in space"—to encompass all reality. One cannot but wonder whether, in that regard, the Hindus may be wiser than we.

of the prevailing scientific approach. One knows by now that the neuronal mechanism associated with visual perception, as Sir Francis Crick reminds us, is designed "to take the picture apart," which is to say that there exists no corresponding neuronal appa-ratus designed to put the picture "back together again."[4] How then is that prodigy accomplished? How does one survey and integrate into a single picture the "on-off" states of a myriad neurons—and do so in a "split second" no less! This is the famous "binding prob-lem" which, as I maintain, actually *admits no solution on the corporeal plane.* As I have noted elsewhere,[5] this "putting together" is in truth effected by what is traditionally termed the *soul* or *anima*, which is able to accomplish this feat precisely by virtue of the fact that it pertains—not to the corporeal—but precisely to the *intermediary* realm, where spatial separation does not exist. It follows that visual perception originates—not on the corporeal—but on the *interme-diary* plane.

I would point out that the binding problem for visual percep-tion is in fact analogous to the measurement problem for quantum mechanics in that it proves to be insoluble under the customary ontological assumptions. It can thus serve likewise to reveal the existence of what may be termed the "next higher" ontological plane: in the measurement problem this is the corporeal, and in the binding problem it is the intermediary.

Having concluded that visual perception originates thus on the intermediary plane, we should not neglect to point out that this ontological fact entails an etiological consequence: inasmuch, namely, as the causality upon which physics is based demands a transmission through space, and that mode of causation is con-sequently no longer operative on the intermediary plane where spatial bounds are transcended, it follows that the causality upon which visual perception is based must in fact be *vertical*. And that in itself explains the current impasse: it follows namely from this etiological fact alone that *all attempts to understand the*

---

4. For a brief introduction to this field of current scientific inquiry I refer to my chapter, "Neurons and Mind," *Science & Myth* (Angelico Press, 2012).

5. *Physics and Vertical Causation* (Angelico Press, 2019), pp. 37-41.

phenomenon of visual perception by way of a physics-based science are bound to fail.

∾

Visual perception, we have said, originates on the intermediary plane. Yet what we normally perceive are objects situated on the corporeal! By what means, then, can this ontological gap be bridged? The answer, I propose, is to be found in the Sānkhya doctrine: in the relation between the *tanmātras* and *bhūtas*, which establishes a bridge or connecting-link between the two ontological tiers. The point is that we perceive the corporeal *bhūtas* by way of the corresponding *tanmātras* pre-existing on the intermediary plane. In a word, *corporeal* qualities derive from the *intermediary* plane, where they pre-exist in a "*tanmātric*" form that renders them perceptible.

This brings us to a pivotal recognition: the distinction between *tanmātras* and *bhūtas* is indicative of *a pre-existence of corporeal objects on the intermediary plane.* And actually this should come as no surprise: as every good Platonist knows, the corporeal object pre-exists even on the aeviternal plane—for the simple reason that *being* is not to be found in the temporal realm. What differentiates the corporeal from the intermediary realm is not being or essence—which is found in neither—but simply the imposition of spatial bounds, which apply to the corporeal but not to the intermediary. And I would add that the alchemists of old understood this fact very well: the two fundamental operations of their science—the so-called *solve* and *coagula*—refer after all to the lifting and the imposition, respectively, of these spatial bounds! Let me emphasize—to drive home the point once and for all—that this pre-existence of the corporeal on the intermediary plane is by no means a pre-scientific fantasy: if namely corporeal entities did *not* pre-exist on the intermediary plane, they would *ipso facto* be imperceptible.

I am persuaded, moreover, that this immemorial doctrine—incomprehensible to the modern mind though it may be—rests on an empirical basis no less rigorous in its own way than that upon which our present empirical sciences are based. The decisive point is that in the traditional sciences it is the scientist himself that serves in a sense as the "instrument"—not of measurement, to be sure—but of *sight*, precisely. What stands at issue, moreover, is a "seeing" capable of perceiving corporeal objects on the intermediary plane: that is to say, ontologically *prior* to their corporeal manifestation. Given furthermore that even ordinary visual perception, as we have noted, originates in truth on the intermediary plane, the possibility of such a "higher" visual perception is hardly in doubt. It is however beyond question, finally, that a science based directly upon visual perception on the intermediary plane is inherently more powerful than our own, which cannot so much as detect the existence of that intermediary realm. As a rule, the higher the ontological plane on which we know, the more powerful the knowing.

But getting back to the Sānkhya, let me note that this traditional science entails, in a way, a refinement of the Aristotelian doctrine of forms. For whereas Aristotle accounts for perception by the fact that a given form can reside simultaneously in the perceived object and its percipient, Sānkhya takes into account the corresponding ontological discrepancy: the fact that the object is corporeal whereas the percipient pertains to the intermediary domain. According to Sānkhya doctrine thus, Aristotle is identifying *bhūtas* and *tanmātras*—which is of course in a way legitimate, inasmuch as what is identified are the gross and subtle manifestations of one and the same quality.

Returning to the problem of (ordinary!) visual perception: having concluded that this perception *originates* on the intermediary plane, we need now to account for the fact that it *terminates* on the corporeal. It is needful, therefore, to distinguish between what may be termed *ontological* and *intentional*[6] visual perception: between what is actually given in that primary perceptual act, and what is intentionally perceived. And as every phenomenologist knows full well, the two are by no means the same. In the words of Henri Bortoft, what is initially perceived is a "whole which is no-thing,"

---

6. In the sense of "intentionality" as the term is used by phenomenologists.

and is therefore as a rule dismissed as "nothing" and replaced with "a world of things,"[7] leaving the percipient with "the apparent task of putting them together to make a whole."[8] That "world of things"—which proves to be none other than the corporeal—is the object thus of *intentional* perception.

Suffice it to say, in regard to this secondary mode of visual perception, that it constitutes a complex process involving presumably other senses—the kinesthetic and the sense of touch for example—in addition to the visual, as well as elements of memory. Intentional perception, thus, is something to be learned, normally in infancy. All this activity, moreover, underlying the simple quotidian act of perceiving a mountain, for instance, or a tree, transpires evidently on the intermediary plane. Corporeal being, one might say, presents itself at the very end of the intentional process as that which is perceived.

It needs finally to be noted that where there is an object—in this instance, the corporeal "what" that is visually perceived—there must be a corresponding subject: the "who" that perceives this "what." The perception of corporeal objects is thus the act of a percipient pertaining to the intermediary plane—the *psychic* percipient we will call it.

～

At this juncture a problem presents itself which proves to be no less fundamental and daunting than the measurement and the binding problems. And even as the latter both demand the crossing of an

7. One of the first hurdles, it appears, to be overcome by a Sānkhya apprentice is *not* to dismiss the "no-thing." And clearly, this is to be achieved by an appropriate discipline of meditation.

8. *The Wholeness of Nature* (Lindisfarne Press, 1996), p. 14. Henri Bortoft was a physicist, a student of David Bohm no less, who became fascinated with the approach to natural science pioneered by Johann Wolfgang Goethe. His book is in fact subtitled, "Goethe's Way Toward a Science of Conscious Participation in Nature." It is safe to say that few if any Western savants in post-Enlightenment times have had as profound an insight into the traditional sciences as Goethe.

ontological threshold—the measurement problem from the physical to the corporeal, and the binding problem from the corporeal to the intermediary—so the third demands that we go the rest of the way: from the intermediary to the aeviternal, namely.

One might refer to this third impasse as "the motion witnessing problem": for if the observer or witness were himself "fully subject" to the bound of time, he could not perceive *motion*, could not perceive *events*. For indeed, what such a time-bound percipient would perceive could be no more than a succession of static images: no more than what a "motion picture" camera registers. One arrives thus inevitably at the ontological recognition that *the ultimate* or *authentic* observer—the veritable witness of a changing panorama—must in truth be situated on what we have termed the *aeviternal* plane.

Where then, let us ask first of all, does this leave our so-called psychic percipient, situated as he is on the intermediary plane? It turns out that he does not stand alone, but is inseparably joined to that "aeviternal" witness. And let us not fail to note, once again, that every authentic Platonist could have told us this from the start: as the Master himself says somewhere, "*The soul is partly in time and partly in eternity*." So long, then, as we conceive of the soul as *psyche*, we need to realize that it is complemented by an aeviternal witness, call it *pneuma*, *nous*, or *spiritus*. We need to remember, once again, that man is neither corporeal, nor psycho-corporeal, but incurably *tripartite*.

I find it astounding that the cognitive psychologist James Gibson[9]—that paragon of "no-nonsense" empiricism—came, *on empirical grounds*, as close as one conceivably can to that very conclusion. The gist of what Gibson discovered—in the course of his painstaking research designed to reveal how we visually perceive—is that this perception is based on what he terms "the pickup of invariants" from the ambient optic array. The crucial discovery is that these "invariants" can be recognized as such only in a state of motion: invariance, after all, manifests *in motion* alone. Visual perception constitutes thus an inherently *kinetic* process:

9. We have discussed his "ecological theory of visual perception" in chapter 5.

The eye is never literally fixed. It undergoes a series of miniature movements or microsaccades... Looking is always exploring, even so-called fixation... The visual system hunts for comprehension and clarity. It does not rest until the invariants are attracted.[10]

"Events are perceived," Gibson assures us; and to enable us to comprehend this prodigy, he goes on to point out that "there is no dividing line between the present and the past, between perceiving and remembering."[11] Now this is very deep, and tantamount, I take it, to the ontological fact that "the present is not a part of time" as Aquinas declares. What is it then, that elusive "present"? It proves to be none other than what the Scholastics termed the "*nunc stans*" or "now that stands." And where else, in the final count, can "the perception of events" take place than in this *nunc stans*! Herein then—in this very recognition—resides the key to the enigma: what ultimately enables the perception of events—of *motion*—is the fact that the primary percipient resides in the *nunc stans*, which in truth is none other than *aeviternity*.

The authentic or "ultimate" seer is not after all the psychic percipient, but proves rather to be what in Indian tradition has been referred to from time immemorial as the *Purusha*, the aeviternal witness of "events." But where, then, does this leave the psychic percipient? The two are in fact inseparable: that is to say, the psychic observer derives his power of vision directly from the *Purusha* himself. And yet he knows not that *Purusha*: his gaze, as we have noted, is directed "outwards" into the corporeal realm. One is reminded of the distinction between "soul" and "spirit" which likewise proves difficult to conceive: so much so that, as St. Paul seems to imply, it ultimately requires "the word of God"—"sharper than any two-edged sword"—to accomplish that "dividing asunder of soul and spirit."[12]

～

10. *The Ecological Approach to Visual Perception* (Lawrence Erlbaum Associates, 1986), pp. 220, 222.

11. Ibid., p. 253.

12. Heb. 4:12

The "big picture" is beginning to emerge. Visual perception does not, of course, take place on the corporeal plane. But neither does it commence on the intermediary as our distinction between "ontological" and "intentional" perception might suggest. The point is that what transpires on these two planes—necessary though it be—does not suffice, does not in itself constitute visual perception: it happens that the most essential "component" remains yet to be identified. It emerges namely from the preceding considerations that visual perception *does not*—cannot in fact—*take place within the spatio-temporal realm*! But whereas the *Purusha* proves to be indeed the "ultimate" witness, the process of visual perception entails a psychic or "ego-centric" observer as well.

It is to be noted that this "ego-centric" observer is facing "outwards" towards the corporeal plane,[13] which means, from a Platonist point of vantage, that he is indeed constrained to gaze upon "shadows" as the celebrated myth has it. By the same token, moreover, he is unaware of the *Purusha*, that "ultimate witness" from whom his own power to perceive is derived. There are thus, seemingly, *two* seers: the egocentric and the spiritual, the temporal and the aeviternal; and the disconnect between the two—between that aeviternal *Purusha* and so-and-so gazing upon the petunias in his garden—could not be more extreme.

I would like now to point out that this "break" between the *psychic* and the *aeviternal* components of our integral being—this congenital inability to discern the aeviternal basis of reality—constitutes in fact the mark of what in Judeo-Christian tradition is termed the *Fall*. The fact is that man in his present state has been severed, as it were, from the spiritual component of himself—his "summit" one might say—and reduced to the status of what St. Paul terms a *psychikos anthropos*: a merely *psychic* being "*who knoweth not the things of God because they are spiritually discerned.*"[14] In other words, he has lost the capacity of spiritual vision by which one can know the aeviternal things that prove to be the

---

13. We are referring, of course, to the waking state: in the dream-state, to be sure, the psychic percipient is facing into the intermediary domain, wherein he perceives *subtle* (as distinguished from *corporeal*) objects.

14. 1 Cor. 2:4

ultimately real. His condition is thus indeed that of Plato's "prisoners" constrained to gaze upon shadows projected upon the wall of a cave! One must not however conceive of this Fall as an historical event—as something that came to pass "in time"—which would in a way be "putting the cart before the horse." The point is that the Fall, so far from having transpired at some remote time in the primordial course of history, is rather to be conceived as *the prehistoric Event that caused or initiated "history" as such.*

I wish moreover to correct a widespread misunderstanding by pointing out that, despite unquestionable concordances, the recognition of the Fall as such is manifestly indigenous to the Judeo-Christian domain. Hindu tradition, for instance, clearly recognizes the resultant impairment, but not the Fall as such; the following verse from the Katha Upanishad may perhaps suffice to make this clear: "*The Self-existent Lord,*" the Upanishad declares, "*afflicted the senses so that they look outward and perceive not the inner Self.*"[15] But whereas this acknowledges in the clearest terms that the senses "*look outward and perceive not the inner Self,*" it ascribes the cause not to an infidelity of the *anthropos*—what Christianity refers to as Original Sin—but to "*the Self-existent Lord*" himself. There exists thus a profound and absolutely fundamental discrepancy between the Vedic and the Christian outlook, which tends in various quarters to be systematically overlooked, to the detriment of clarity regarding the most decisive issues.

∾

Let us assume then that religion, in its higher manifestations, is able in some degree to reverse that Fall—to "narrow the gap," as it were, between the *psychikos anthropos* and the spiritual component of our integral being—and that such a restitution constitutes in fact the measure of what is *de jure* termed *sanctification.* The question arises now whether that sanctification—this at least partial "undoing" of the Fall—may impact the powers of visual perception so as to bestow a distinctly "supernatural" capacity to transcend

15. II.1.

not only spatial, but *temporal* bounds? Can it perhaps free the saint from the limitations of the psychic percipient so as to perceive as it were from the vantage of the aeviternal plane? I am persuaded that such is indeed the case.

Let one example suffice: that of Anna Katharina Emmerich (1774-1824), an Augustinian nun whose convent in Germany was closed in 1812 as a result of the Napoleonic incursion, and who subsequently resided in a farm house, confined to a single room and bedridden. The Catholic reader may be interested to know that according to a document attested by a group of physicians, Anna Katharina was in truth a stigmatic. What concerns us is that she spent much of her time in trance states in which, according to her testimony, she visited—in the primary sense of *seeing*—various far-away regions, from Palestine to the Himalayas. What is more, she claimed to have "seen" events pertaining to the distant past, and likewise—though more rarely—to a future century: and that, obviously, is what a *psychikos anthropos* is categorically unable to do.[16]

Much of what Anna Katharina describes as having seen—typically in meticulous detail—are scenes from the Old and New Testament, and in these depictions quite a bit is altogether new. What lends credence to this testimony is the fact that wherever her account—e.g., descriptions of local geography, inclusive of facts the seeress herself could not possibly have known by natural means—could be verified, her reports have invariably proved true down to the least detail. To cite just one example: her depiction of the house in which the Virgin Mary spent the last years of her life—which is located in the foothills above Ephesus, and had not at the time been identified—was discovered on the basis of Anna Katharina's descriptions by a French priest in 1881, long after she and her biographer, Clemens Brentano, had departed from this world.

How, then, can such feats of clairvoyance be explained? Now, so far as seeing at a great distance is concerned, this can in principle

---

16. The interested reader may wish to consult the impressive series of "Emmerich books" brought out recently by Angelico Press.

be understood on the basis that ontological seeing takes place, after all, on the intermediary plane, in which there is in truth no spatial separation at all. And as a matter of fact, it appears that certain individuals—often referred to as "mediums"—are gifted with a preternatural capacity to "visually perceive" distant objects and events. When it comes to seeing into the past or future, on the other hand, the "psychic percipient" is evidently not up to the task. What is called for is the "aeviternal percipient": he alone can see what for us are presently past or future events. It appears thus that Anna Katharina was endowed with a capacity to "perceive from the aeviternal plane," which is to say that she had, in some measure at least, gained access to what, in Judeo-Christian terms, could be termed the "Edenic state." But this should hardly come as a surprise: does not authentic sainthood, as we have noted, entail precisely an access of that kind?

∼

When all is said and done, visual perception remains for us a mystery: it constitutes, after all, an act of the tripartite wholeness we all are, but cannot, *qua psychikos anthropos*, know that we are. In every act of visual perception, thus, *three* ontological strata come perforce into play—even though we are normally aware only of one: i.e., the corporeal. The intermediary plane does also, as we have noted, come into play—but we are not as a rule conscious of that realm. When Aldous Huxley, following a practice of certain American Indian tribes, imbibed mescaline, he entered into the lower reaches of the intermediary, and segments of our population have been doing much the same ever since. Yet pending the restoration of the traditional sciences, that domain is bound to remain for us inherently a closed book. And finally, when it comes to the aeviternal realm, who among us has had so much as a fleeting glimpse of that supreme cosmic sphere!

11

GNOSTICISM TODAY

BASED UPON the almost universally overlooked distinction between scientific fact and scientistic belief, I have had much to say in recent years concerning the contemporary worldview: it happens that we have been collectively duped. So far, namely, from standing upon solid scientific ground as we are taught to believe, that *Weltanschauung* derives in truth from a very different source: the fact is that in the hallowed name of Science we have fallen prey *en masse* to an *ideology*. It behooves us then, in the wake of this recognition, to take the next step: namely, to investigate that underlying ideology itself. There is in fact no other way of attaining authentic discernment in regard to that sovereign worldview which seems nowadays to impose itself on just about everyone, from heads of church and state to college freshmen, at the very least.

We propose now to examine that "ideology" as such; and in so doing we shall come to understand science in a new key. We shall thus discover, in particular, the source of its anti-religious thrust: the fact that the scientistic animus *against* religion is itself something "religious"—even as a disclaimer of "politics" may be political. In the end, what underlies and secretly "drives" the modern world proves thus to be a kind of religion: a religion "turned upside-down" one might say, which in fact reduces to that ancient and protean counter-religion known to historians as *Gnosticism*.

Now it happens that, more than three decades ago, I published an article on that subject[1] in which I argued that this archaic heresy has morphed—as in fact it has ever been wont to do—into

1. "Gnosticism Today," *Homiletic & Pastoral Review*, March 1988.

yet another version of itself, and that under this new guise it has infiltrated Western civilization. This intrusion has been in progress more or less since the Enlightenment, and appears to have succeeded brilliantly in subverting, first the educational institutions—from universities down to elementary schools—and from thence extending its reach throughout the culture, at last infiltrating even the seminaries. We seem to be nearing the Gnostic tipping point. The fact is that the proponents of Gnosticism—or neo-Gnosticism, more precisely—are presently waging all out warfare to secure hegemony over Western civilization. And speaking now as a Christian, I would add that nothing short of authentic religion can stand against that Gnostic onslaught: *"For we wrestle not against flesh and blood, but against principalities, against powers, against the rulers of the darkness of this world, against spiritual wickedness in high places"* as St. Paul apprises us.[2] Let those then who have "ears to hear" reflect upon these words! Being "reasonable"—or ever so "rational"—will not suffice: neither bourgeois "good sense" nor credentialed "expertise" will make so much as a dent against the Gnostic onslaught.

Notwithstanding what we have been taught to believe in colleges and universities, we are all in essence "religious": made *"in the image and likeness of God"* no less, as the Good Book informs us, we *cannot*, actually, choose *not* to be. Our only option, it turns out, pertains to the *kind* of religion we shall make our own. And there are, at bottom, only two choices: Christ speaks doubtless for true religion at large in declaring: *"He that is not with me is against me."*[3] Ultimately we must choose to which of "the two cities"[4] we shall henceforth belong; and one way or the other, each of us *will* make that choice.

～

Before one can speak of "Gnosticism" as a contemporary movement or trend, one needs to disengage the definitive features of this

2. Eph. 6:12
3. Matt. 12:30
4. The metaphor derives of course from St. Augustine's justly famous treatise *Civitas Dei*.

ideology from the welter of ideas—for the most part exceedingly weird—to be encountered in the ancient Gnostic texts. To be sure, the curious doctrines of Simon Magus, of Marcion or Valentinus, are dead and gone, unless perchance they have been resurrected in our day by some outlandish cult. What primarily interests us, in any case, is the astonishing fact that quintessential Gnosticism has reasserted itself in post-medieval times—not in the form of some exotic mysticism—but precisely *in and through the mainstream of contemporary culture.*

Ancient Gnosticism, as one knows, did not present itself as a well-articulated teaching or unified doctrine: the very opposite, actually, has ever been the case. In fact, one of the most striking characteristics of the erstwhile Gnostic schools is their extreme syncretistic tendency; as Hans Jonas points out: "The Gnostic systems compounded everything—oriental mythologies, astrological doctrines, Iranian theology, elements of Jewish tradition, whether Biblical, rabbinical, or occult, Christian salvation eschatology, Platonic terms and concepts..."[5] And as if that were not enough, individual Gnostic gurus were by no means reticent to contribute novelties of their own; as St. Irenaeus informs us: "Every day every one of them invents something new."[6] Small wonder that within the domain of Judeo-Christian Gnosticism alone, scholars have counted as many as thirty different speculative schools.[7]

Yet even so, there exist of course common elements, failing which one could hardly speak of "Gnosticism" at all; and I would contend that the most distinctive of these features, which invariably plays a key role in the economy of Gnostic thought, is what may be termed *the Gnostic devaluation of the cosmos.* It is to be noted that this tenet constitutes the direct negation of a corresponding Christian belief: the contention, namely, that the world was created by a beneficent God, and is itself inherently *good.* According to this Christian belief—which apparently was shared more or less by the major philosophical schools of antiquity—the uncorrupted

---

5. *The Gnostic Religion* (Beacon Press, 1963), p. 25.
6. *Adversus Haereses* I.18.1.
7. Joseph Lortz, *History of the Church* (Bruce, 1939), p. 65.

cosmos is in truth a masterpiece, as perfect as anything not itself divine can be. What is more, it is said to be actually a theophany, a kind of reflection or image of God himself; for as St. Paul declares: "*The invisible things of him, from the creation of the world, are clearly seen, being understood from the things that are made, even his eternal power and Godhead.*"[8] And if it happens that we ourselves do not behold these "*invisible things,*" it is ultimately *we*—our "*vain imaginations*" and "*foolish heart*"[9]—that stand at fault.

Now this is precisely what the Gnostic can never admit: so far from constituting fallen creatures in an inherently good and ultimately *theophanic* world, the Gnostic gurus proclaim us to be in effect quasi-divine beings, who through no fault of their own have been cast into an alien, cruel, and perfectly senseless world. And if it happens—as all too often it does—that we ourselves engage in vile and despicable acts, it is invariably the world at large that is at fault: it is this enslavement, this unjust compulsion that cramps our godly style! One is vividly reminded of Rousseau's "noble savage," and of Freudian neurosis as the effect of externally mandated inhibitions. And this constitutes in fact a central dogma on which Gnostics of every stripe invariably insist: in the final count it is always the world—the cosmos with its inexorable law, its *heirmarmene*—and not our "*foolish heart*"— that holds us in chains.

But let us note what this entails. If all our misery derives indeed from the external world into which we have been undeservedly cast, then it behooves us to revolt: to seek by whatever means to liberate ourselves from the shackles of this world. Under such auspices one is prone to believe that nothing more is required for the attainment of unmitigated bliss than to divest oneself radically of external constraints. Such then is the second fundamental dogma of classical Gnosticism: the notion that the *summum bonum* is to be achieved by a radical separation from the conditions imposed upon us by our terrestrial environment. And the third fundamental dogma of ancient Gnosticism is that this liberating act is to

8. Rom. 1:20
9. Rom. 1:21

be realized as a "mystic journey," a mysterious flight into "higher worlds."

A word of warning is however called for at this point: we must not be too quick to perceive a "Gnostic journey" in every ancient mysticism we chance upon. When Plotinus, for instance, speaks of a "flight of the Alone to the Alone," it does not instantly follow that he is a Gnostic. It needs thus to be realized that there exist in truth different kinds of "mystic flight"—ranging from that of a Plotinus all the way down to the psychedelic "high" of a hippie—which so far from being commensurate, are in truth opposed, somewhat as heaven is to hell. In this domain—more than in any other—it behooves us to exercise caution and prudent restraint: the stakes are high.

Getting back to Gnosticism: once that "mystic flight" has come to be perceived as the liberating act *par excellence*, there remains but one question: by what means or *modus operandi*, namely, is that crucial exodus to be effected? And it is in fact from their generic answer to this question that the Gnostic sects derive their common designation: "the one thing needful" for the attainment of that Gnostic Passover, namely, is said to be *gnosis*.

Now "*gnosis*" is simply a Greek word meaning "knowledge," which for thousands of years has been used by philosophers and theologians to refer in particular to the "higher"—or "supreme"—knowledge attainable through the quest of God. What happened is that the Gnostic gurus usurped that term by turning "*gnosis*" into the *modus operandi* of their own Gnostic Passover, a practice which ended up giving the word a bad name. Never mind the fact that St. Paul speaks of a high and venerable *gnosis*,[10] and that Christ himself rebukes the Pharisees for having "*taken away the keys of gnosis*" as we learn from the Greek text of Luke 11:52! Yet, even so, the offending term has in effect been demoted within the academic world to its Gnostic counterfeit, with the result

---

10. St. Paul distinguishes sharply between authentic *gnosis*—which bestows a knowledge of God—and what he terms "*pseudonymou gnoseos*" or "*gnosis* falsely so named" (1 Tim. 6:20). It hardly needs saying that for St. Paul the Gnostic gnosis was indeed a prime example of such a "*pseudonymou gnoseos.*"

that authentic *gnosis* is hardly ever mentioned anymore "in polite society."[11]

As to the role of *gnosis* in the Gnostic enterprise, what actually stands at issue is not *gnosis* as such—which in truth the Gnostics do not possess—but its counterfeit in the form of the "Gnostic secret," by means of which that *gnosis* is reputedly brought into play. In brief, Gnosticism comprises *four* essential elements: first, a radical devaluation of the cosmos; secondly, a grievance against an oppressive status quo; thirdly, an eschatological aspiration taking the form of a "mystic journey" as an escape from that status quo; and finally, the stipulated possession of the "Gnostic secret" which supposedly enables that liberating act.[12]

It is to be emphasized that belief in a "liberating *gnosis*" does not—all by itself—constitute Gnosticism. If such were the case, it could indeed be claimed that religion *per se* is Gnosticism: for where does one encounter a *bona fide* religious tradition which does not, in one way or another, allude to a sovereign *gnosis*? Despite its conspicuous emphasis on *love*—on *agape*—Christianity moreover is no exception in that regard: *"Ye shall know the truth, and the truth shall make you free"* the Savior himself declares![13] Think of it, *"the truth shall make you free"*: what indeed could be more "Gnostic" than that! If these words had been discovered on some ancient

---

11. It may be worth pointing out that the Sanskrit word *jñāna* derives from the same Indo-European root as gnosis. What Hindus term *jñāna* yoga is therefore literally the way or *"yoga"* of *gnosis*: authentic *gnosis* namely, as distinguished from its Gnostic counterfeit.

12. There exists at least a superficial resemblance here with the "four noble truths" of the original or *hinayāna* Buddhism: i.e., "the existence of suffering, the cause of suffering, the cessation of suffering, and the way leading to the cessation of suffering." Actually however, Buddhism in all of its forms constitutes the very opposite of Gnosticism: no respectable Buddhist, namely, has ever blamed "the universe" for his own miseries. Buddhism puts the blame for suffering squarely on our own ignorance and inordinate desires: it is we—and not the universe—who are at fault! Which entails, moreover, that we too can remedy that condition in which we presently find ourselves. What in pristine Buddhism replaces the "mystic flight" of the Gnostic is a radical separation—not from the world as such—but from our desires and cravings.

13. John 8:32

tablet or papyrus, would not our *periti* conclude at once that it is a Gnostic text?

~

It is to be noted that the Gnostic deprecation of the cosmos leads naturally to a rejection of perennial norms and traditional admonitions. First of all it causes the Gnostic to be ever the nonconformist, the spoiler who typically delights in disrespecting whatever is normally perceived to be venerable, let alone sacred. As a rule, the Gnostic finds it hard to concur on anything even with his fellow Gnostics; and when it comes to the non-Gnostic portion of mankind, his sympathies tend to run exceedingly thin: any deviation from Gnostic dissent or disbelief is apt to provoke contemptuous rejection even where his own are concerned. The Gnostic's antipathy to the cosmic order extends into the cultural sphere: he is disposed against everything, in fact, which presents itself as a given, a status quo of whatever kind. In a word, the Gnostic is a born revolutionary, a creature of *ressentiment*: the postulate of unmerited misery, it appears, is apt to arouse antagonism and bitterness in even the gentlest soul.

The irony, however, is that this *ressentiment* itself obstructs the realization of the Gnostic *telos*, its quest of world-transcendence, to the point of rendering such a "liberation" unattainable in principle: where there is so much as a trace of anger or ill will, there *can be no* emancipation of any kind—what to speak of *gnosis*. Authentic world-transcendence calls in truth for the very opposite: "*Take my yoke upon you,*" Christ declares, "*and learn of me; for I am meek and lowly of heart: and ye shall find rest unto your souls.*"[14] It appears however that the Gnostic relishes the very opposite: that in fact he is literally inverting the teaching of Christ. Let us understand the point as sharply as we can: if Christ had said only "*ye shall know the truth, and the truth shall make you free,*" he might conceivably have been a Gnostic; but he also said: "*Love thine enemies, do good to them*

14. Matt. 11:29

that persecute you"[15]—and no Gnostic could ever bring these words across his lips.

～

To the casual observer nothing, to be sure, could seem more out-landish and "dated" than the speculations of the ancient Gnostic sects. It needs first of all to be recalled that with the advent of the Renaissance, the dominant interest of Western man began to shift: from God and transcendence to the exploration and eventual mastery, namely, of the visible world. It was a time of transition, when rationalism and skepticism regarding the "higher worlds" began to afflict the European intelligentsia, preparing the way for the Enlightenment. The medieval propensity for theological speculation was fast falling out of vogue, while in the universities the ground was being prepared for the scientific revolution soon to get under way. It would seem that by the fifteenth century or thereabouts, a visible disenchantment with the mystical quest as such was beginning to manifest. A growing contingent among the emergent ranks seemed ready and eager to give themselves to far more tangible pursuits. A new *Zeitgeist* was beginning to emerge: it seems that rarely in history had this world appeared quite so fair and inviting to mortal eyes. A brave new breed of men—far more interested in building empires than in mystic journeying of any kind—had sprung up, and was fast taking over the helm.

And so, by the time of the Enlightenment, something altogether unprecedented had taken place: as Huston Smith points out, it happens namely that "the modern West is the first society to view the physical world as a closed system."[16] Never before, it seems, had mankind in general—and the intellectual elite especially—forgot-ten so completely the existence and function of what, from time immemorial, mankind had conceived as "higher spheres."

Under such auspices Gnosticism is still of course conceivable as a sub-culture or counter-culture in opposition to the prevailing

15. Matt. 5:44
16. *Forgotten Truth* (Harper & Row, 1977), p. 96.

Zeitgeist. But how—in a civilization that has abjured transcendence—could there be a Gnosticism *within* the cultural mainstream? How indeed could one speak of a "flight from this world" when this world had in effect become all that exists? To do so the ancient doctrines needed evidently to be interpreted in a new key: a contemporary Gnosticism, to be conceivable, calls evidently for a new hermeneutic, a new way, that is, to envision the Gnostic "flight." And so in fact it does. As Eric Voegelin points out, the requisite reinterpretation was in fact achieved through what he terms "the immanentization of the Eschaton": in place of an Eschaton transcending the ontological confines of this sense-perceived world, the modern Gnostic envisions an End to be realized on this very ontological plane, within the confines of human history. To speak in geometric terms, there has been a rotation of 90 degrees, turning the "above" into the "ahead": eschatology has thus been transformed into a kind of social activism, a "social engineering" if you will.

To complete the picture, it is to be noted that even so a hunger to transcend what the poet terms "this narrow world" remains in the neo-Gnostic as well, and drives him as a rule to seek a taste, at least, of transcendence, here and now, by whatever means remain. And let us observe that there are in fact two such means within easy reach, one being psychedelic drugs, the other being sex, which has ever served many as "the poor man's mysticism."

∽

It is perhaps surprising that the key notion of neo-Gnosticism was actually prepared for by Christianity itself. It happens, namely, that the Christian worldview is the first to endow the cosmos as such with a *telos*, an "end" in both senses of the term. Under the aegis of Christianity history came namely to be viewed no longer as an endlessly recurrent series of cycles, but as a directed movement in quest of a final encounter with Christ in what is termed the Parousia, at which point history as such will come to an end. Just one more step, formally speaking, was needed to transition from this Christian eschatology to the neo-Gnostic Eschaton: given that history has an End, an Eschaton, *that End needed to be conceived as taking place within the bounds of history itself, as its final phase.*

But for this too, Christianity had in a way prepared the ground: for it happens that the conspicuous clarity of the early Christian teaching was eventually compromised by chiliastic speculations which tended in effect to immanentize the Eschaton by postulating an imagined millenary reign of Christ. Already in the fourth century St. Augustine had labored to put an end to such notions, and succeeded in settling the matter theologically for the Church. Yet not only did chiliasm survive, but during the latter half of the Middle Ages gave rise to a movement of some magnitude. From the time of Joachim de Fiore, it seems, up to the Renaissance, Europe was rife with millenary speculations, which at times erupted in frenzied movements of unrest. And in all these manifestations of what historians are wont to call "the pursuit of the millennium," we encounter one and the same presiding idea: a belief, namely, in a collective salvation to be realized here on Earth through a radical transformation of some kind.

To be sure, medieval chiliasm presented itself in Christian colors and was as a rule perceived by its votaries as the authentic Christianity no less. Yet despite superficial appearances and vehement protestations to the contrary, heretical Christianity is never in fact Christianity at all. As has been noted time and again: deny one dogma—one seeming fine point of fundamental theology—and you have implicitly denied all the rest. And this explains why chiliasm—once it had immanentized the Parousia—could readily shed its Christian garb and morph into a medley of anti-Christian sects.

It is thus by no means unlikely that the key notion of the neo-Gnostic Eschaton—the seductive panorama of a terrestrial and futuristic salvation—was bequeathed to the modern world by the millenary movement of medieval Christendom. These fantasies may well be what enabled classical Gnosticism to transplant itself into the modern age: the masterstroke which permits the post-medieval Gnostic to "fly into a higher world" when in fact there are no more "higher worlds" to fly into! Let us be clear on this crucial issue: the Gnostic revelation for the New

Age asserts that there *is no* "higher world"—nor shall there ever be—*unless it be created by man himself.* The "heavens" and "paradises" of religion are but a fantasy—until, that is, they are realized through the labors and ingenuity of man. And God himself—so the neo-Gnostic declares—is finally but a premonition of the coming Superman.

∽

But let us remind ourselves that one "myth" does not make a Gnosticism: there are invariably *four* components, as we have seen, the first being what may be termed "the Gnostic devaluation" of the cosmos. Now this too may strike us as antiquated: how after all could the cosmos be "devalued" when it has in fact been elevated to the prime reality? Yet, on closer examination, one sees that this neo-Gnostic postulate is itself a "devaluation of the cosmos," and in fact the most radical at that: for by negating the transcendent and aeviternal nature of *being* as such, every order of existence has been reduced to the status of a contingency, a mere "accident" one might say. And once *being* has been thus "decapitated"—to use Eric Voegelin's marvelous term—everything, both in the natural and the human realm, has lost its sanction, its legitimacy, and above all, its true meaning. *The denial of transcendence constitutes*—in the final count—*the ultimate devaluation of the cosmos.*

Now this denial—this "decapitation of being"—is at bottom the Gnostic "murder" of God: the very "mystery" Nietzsche had his eye upon when he announced the "death" of God. Yet needless to say, this is a "murder" that can only be consummated "speculatively"—in the mind and heart of man—and even that proves not to be an easy task. The enterprise has in fact challenged the brightest minds of Europe and America for about the past four centuries: at the close of the Middle Ages mankind had not yet by any means "progressed" that far! Admittedly, incipient neo-Gnosticism was already in possession of its futuristic Eschaton: had presumably inherited that pivotal notion from heretical Christian sects. The neo-Gnostic devaluation of the cosmos, on the other hand, had as yet barely begun. Nor had anyone yet revealed to mankind the "neo-Gnostic keys" that will one day supposedly empower mankind to enter the proffered Paradise.

Yet, what was missing still was ere long to be supplied. It arrived in stages: one by one the Gnostic gurus made their appearance. There was Jean-Jacques Rousseau, Voltaire, and the tribe of Encyclopedists; after which came Hegel, Marx, and Nietzsche, followed by a host of lesser luminaries too numerous to name. The fact is that the history of neo-Gnosticism all but coincides with the intellectual history of the modern West: for as Eric Voegelin points out,[17] *the re-emergence of Gnosticism constitutes the very essence of modernity.*

∿

One cannot but concur with Carl Jung that Gnosticism is in fact the counter-position to Christianity. To the extent, moreover, that belief in the God of Christianity wanes, it is replaced inevitably by an allegiance to the neo-Gnostic Eschaton: *futurity replaces transcendence*—that is the point. As Voegelin observes:

> Gnostic speculation overcame the uncertainty of faith by receding from transcendence and endowing man and his intramundane range of action with the meaning of eschatological fulfillment. In the measure in which this immanentization progressed experientially, civilizational activity became a mystical work of self-salvation.[18]

Golden words, which should be inscribed over the portal of every university and every seminary in the land! So far from reducing to an arcane and outmoded superstition, Gnosticism in its resurrected form is in truth the counter-religion to Christianity, which in our day appears to have "tipped the scale" to become the presiding ideology in Western civilization.

17. *The New Science of Politics* (University of Chicago Press, 1952), p. 126.
18. Ibid., p. 129.

As Voegelin observes, the *telos* of contemporary Western civilization in its "progressive" modes is derived from the resurrected Gnosticism. By pretending to be "scientific" and value-free, the latter can disseminate its pseudo-religious and anti-Christian credo with impunity while denying authentic religion access to the public square. Having succeeded brilliantly in transforming schools and universities at large into neo-Gnostic citadels dedicated to the mystical work of self-salvation, it has infiltrated just about every civic institution in the land.

The almost universally unrecognized and unsurmised fact is that beneath the contemporary veneer of scientific enlightenment and pragmatic sobriety there lurks a mysticism. The very terms we cherish the most—such as "progress," "freedom," or "science"—have acquired a kind of mystical ring; as Martin Lings points out, they have become "words to conjure with," enchanted vocables "at the utterance of which multitudes of souls fall prostrate in submental adoration."[19] Behind these contemporary mantras one senses the magic power of the New Eschaton: the Gnostic Good News of "self-salvation" which today speeds millions on their mystic way.

But let us remind ourselves that every Gnosticism requires—not just one—but four legs to stand upon. The Good News, in particular, of which we speak, would not make the slightest impact upon anyone if it were not supported by what has been termed "the speculative decapitation of being." And this entails that a second myth is called for to shore up the first: a cosmological myth, thus, is needed to complement the eschatological.

≈

The presiding cosmological myth of the neo-Gnostic movement is doubtless the dogma of evolution, beginning with the Darwinist version thereof. As has often enough been pointed out, Darwinist evolution was never a *bona fide*—let alone a viable—scientific theory, and could have been proposed only at a time when knowledge

19. *Ancient Beliefs and Modern Superstitions* (Perennial Books, 1965), p. 45.

in the relevant domains of science—biology, biochemistry, and geology mainly—was yet in its infancy. The fact is that if Darwinism were in truth a scientific theory, it would have long ago joined "phlogiston" and other such notions as a conjecture which proved to be untenable. But as I say, Darwinism has been from the start an ideological postulate, dressed up in scientific garb: the fact that it has been doggedly retained in the face of Himalayan counter-evidence itself suffices to substantiate that claim. The doctrine of evolution has swept the world, I say—not, to be sure, on the strength of scientific merit—but precisely in its capacity as *the presiding neo-Gnostic myth.*

Darwinism affirms in effect that living creatures create themselves, which constitutes an inherently *metaphysical* claim. A century later, moreover, that Darwinist hypothesis was in a way universalized in the Einsteinian astrophysical cosmology, which is likewise quintessentially evolutionist; and I might add that meanwhile this astrophysical theory too has in turn been disproved.[20] The respective evolutionist visions fit together, moreover, to yield a *Weltanschauung* which conceives of the universe as a single self-creating process, a notion pivotal evidently to the neo-Gnostic synthesis: for only under the banner of "self-*creation*" can the Good News of "self-*salvation*" be seriously entertained.

No one however has brought to light the neo-Gnostic vistas implicit in the evolutionist myth more poignantly than Teilhard de Chardin, who seems to have pursued that notion into its utmost depth. As Hans Urs von Balthasar observed, "Evolution" came to be, for Teilhard, "the only category of thought," swallowing even the concept of God in the process.[21] The Teilhardian cosmogenetic tale begins with a primordial cosmos in which nothing "recognizable" yet exists. Gradually entities corresponding to subatomic particles emerge, which combine to form more and more complex conglomerates. When this complexity reaches a certain threshold, "consciousness" is said to emerge, and in due time this evolutive

20. I have argued the point in *Physics and Vertical Causation* (Angelico Press, 2019), pp. 68-71.
21. "Die Spiritualität Teilhards de Chardin," *Wort und Wahrheit*, XVII (1963), p. 344.

process gives rise to the biosphere. The production of new living entities does not however end there, but continues supposedly on a social, technological, and even political level to form "super-organisms." The grand new idea is that evolution does not terminate at the present level, but continues its "upward" course to bring into being eventually a single super-organism, which Teilhard proceeds to identify as the futuristic Eschaton.

It is all rather simple: if atoms have aggregated themselves into molecules, and molecules into living organisms of every description, then why should not mankind aggregate itself eventually right into the New Jerusalem? The Mystical Body of Christ, Teilhard assures us, is being formed before our very eyes through the exploits of technology and the formation of political super-states.[22] And this evolutive process is said to extend even into the astrophysical domain in the form of a vast convergence to what Teilhard terms Point Omega, and takes to be the fullness of Christ himself! Never mind that "big bang" astrophysics affirms precisely the opposite: that instead of converging to anything, the universe is actually flying apart in all directions at the rate of almost 300,000 kilometers per second—not that this matters any more, now that "big bang" astrophysics itself has fallen.

Yet no matter how unfounded and wrong-headed the Teilhardian "science-fiction theology" proves to be, its impact—especially upon the Catholic Church—has been immense to the point of ushering in a new era. What stands at issue, I suggest, is the seductive power of Gnosticism, the perennial counter-religion: no wonder millions have been duped! Remember: *"we wrestle not against flesh and blood."*

∾

A key notion of the Teilhardian theory is his so-called "law of complexity/consciousness," which in truth reduces to the gratuitous

---

22. Think of it: to perceive a connection between the European Union say, and the Mystical Body of Christ—to dream this up does take genius of a kind!

assumption that—somehow—consciousness "emerges" out of complexity. Yet vacuous and chimerical as that notion proves to be from a scientific or a philosophical point of vantage, this so-called law plays a pivotal role in the Teilhardian doctrine. Teilhard himself, in his *magnum opus* entitled *The Phenomenon of Man*, informs us at the outset that all the rest of this, his premier treatise, is nothing but the application of this single "great law."[23] What is it, then, about that vacuous so-called "law" that renders it central to the point that all the rest of Teilhard's voluminous opus flows, as it were, from that one tenet? The answer—once recognized—proves to be obvious in the extreme: what else, after all, can Teilhard's "law of complexity/consciousness" be than *the neo-Gnostic "secret" of the Teilhardian enterprise!* To grasp what stands at issue, let us recall that Gnosticism, from the most ancient times, has ever been focused upon one thing: its own version, namely, of that "mystic flight into a higher world." The point is that Teilhard's so-called "law of complexity/consciousness" constitutes the key to the consummation of that enterprise. What this law proclaims, in plain words, is that through the pooling of human resources—the formation of a "super-organism" as Teilhard has it—man is enabled to take control of his own evolution: *can in fact build up the Mystical Body of Christ*, which has now morphed into the neo-Gnostic Superman.

Given the enormity of the Teilhardian influence and the centrality of his role in restructuring the Catholic Church, let us look a bit closer into this remarkable claim.[24] Once one has glimpsed its neo-Gnostic core, just about everything Teilhard proclaims reveals a hitherto unsurmised perversity. Consider for instance his response to the first detonation of an atomic device: for Teilhard this was not just an outstanding feat of science, but a milestone in the spiritual ascent of man "pointing the way to his omnipotence"![25] Continuing to muse upon that event, he goes on to declare—in unmistakably

23. *The Phenomenon of Man* (Harper & Row, 1965), p. 61.
24. A thorough treatment of this subject may be found in my monograph, *Theistic Evolution: The Teilhardian Heresy* (Angelico Press, 2012).
25. *The Future of Mankind* (Harper & Row, 1964), p. 148.

prophetic tones: "In laying hands on the very essence of matter,[26] we have disclosed to human existence a supreme purpose: the purpose of pursuing ever further, to the very end, the forces of Life."[27] The explosion of an atom bomb, it appears, was for him but a step in a gargantuan enterprise, "the first bite at the fruit of great discovery": the "discovery," it appears, of how at last we may make good the ancient promise "*Ye shall be as gods*"! Let us see, then, how Teilhard envisions the realization of this tantalizing prospect. Referring back to the atom bomb, he goes on:

> Was it not simply the first act, even a mere prelude, in a series of fantastic events which, having afforded us access to the heart of the atom, would lead us on to overthrow, one by one, the many other strongholds, which science is already besieging.[28] The vitalization of matter by the creation of supermolecules. The remodeling of human organisms by means of hormones. Control of heredity and sex by manipulation of genes and chromosomes. The readjustment and internal liberation of our souls by direct action upon springs gradually brought to light by psychoanalysis.[29] The arousing and harnessing of the unfathomable intellectual and effective powers still latent in the human mass...[30]

Is it necessary to point out that this is the purest neo-Gnosticism? One is amazed that even such an utterly odious and overtly Promethean outburst as this did not forewarn the faithful as they thronged adoringly around the Master by the millions!

26. As every true metaphysician knows, it is in fact "the very essence of *matter*" not in truth to have any *essence* at all. Placing "essence" in *matter* rather than in *spirit* (the opposite pole) is in fact indicative of the Gnostic inversion.

27. Op. cit., p. 151.

28. Let us note that some eight decades after this prediction was made, it appears that most of these "strongholds" stand safe and sound. As we have noted in chapter 4, the all-conquering march of physical science is not proceeding according to plan.

29. On the nature and occult dangers of psychoanalysis I refer to chapter 5 in *Cosmos and Transcendence* (Angelico Press, 2008).

30. Op. cit., p. 149.

Thus we arrive, finally, at the conclusion of our reflections—at the moral, so to speak, of our story—which reduces to this: *we find ourselves today, both outside and inside the Church, in a predominantly Gnostic environment*, and no Saint Irenaeus has yet appeared to apprise us, with spiritual might, of this ominous fact. The onus is therefore upon each and every one of us to exercise, in the highest degree possible, what our forebears termed "discernment of spirits." Remember, what stands at issue is the salvation of our soul: "*For there shall arise false Christs, and false prophets, and shall show great signs and wonders, insomuch that, if it were possible, they shall deceive the very elect.*"[31]

---

31. Matt. 24:24. To which one might add that so far as we ordinary mortals can tell, the "*deceiving of the very elect*" appears to be in full progress as we speak.

12

BEYOND THE
TRIPARTITE COSMOS

HAVING REJECTED the neo-Gnostic immanentization of the
Eschaton as the imposture it is, let us say a few words regarding the Christian Eschaton, known as the Kingdom of Heaven
or the Kingdom of God. Christianity teaches that this Kingdom
will manifest itself to mankind at the Parousia or Second Coming
of Christ—not neo-Gnostically as the final phase of history, to be
sure—but, on the contrary: as *the End of history* in both senses of
the term. Now this teaching has always been difficult, and in our
day has been grossly misconceived even by theologians of repute.
I wish therefore to clarify this crucial issue, to shed light on what
may aptly be termed the *metaphysics* of the Christian Eschaton.[1]

It is a fact that the eschatological teachings of Christ, as
recorded in the Gospels, contain passages strongly suggestive of
an imminent Parousia or "end of the world." One need but recall
the very first words of Jesus in the Gospel according to St. Mark:
"*The time is fulfilled and the kingdom of God is at hand: repent ye,
and believe the gospel.*"[2] Who among the contemporaries of Jesus,
hearing these words, could have surmised that two thousand years
later humanity would still be awaiting that Kingdom! Or recall the
words of Christ to His disciples as He sends them forth to preach
the Good News: "*But when they persecute you in this city, flee ye into
another: amen I say unto you, ye shall not have gone over the cities of*

---

1. I will draw on my erstwhile article "Eschatological Imminence and Biblical Inerrancy," published in the *Homiletic & Pastoral Review*, May 1987.
   2. Mark 1:15

Israel, till the Son of man be come."[3] Or Matthew 16:28, where Jesus tells the disciples that "*the Son of man shall come in the glory of the Father with his angels,*" following which He goes on to proclaim: "*Amen I say unto you, there be some standing here which shall not taste death till they see the Son of man coming in His kingdom.*" And to give perhaps the most striking example: in the eschatological discourse of Matthew 24-25, after speaking at length of the Parousia, He declares: "*Amen I say unto you, this generation shall not pass till all these things are fulfilled.*"

Such logia, to be sure, are profoundly perplexing. But must we agree with Hans Küng that "they defy any attempt to make them innocuous"? Must we concur when he goes on to say that "Jesus and to some extent the primitive Church speaking with him, and clearly also the Apostle Paul, reckoned with the advent of God in their lifetime"? Küng may of course be right in his claim that "on this, it seems, the leading exegetes are largely in agreement"[4]—but what exactly does that prove?

Well, in the first place it proves that "the leading exegetes" have departed from Catholic tradition: no Father or Doctor of the Church, surely, would have countenanced the tenet that "Jesus, like that whole apocalyptic generation, expected the kingdom of God—the kingdom of justice, freedom, joy and peace—in the *immediate future.*" I will mention in passing that such an interpretation has in fact been contradicted in a number of papal edicts, as well as in judgments issued by the Pontifical Biblical Commission. In *Lamentabili Sane*, for example, Pius X condemns the proposition that "it is impossible to reconcile the natural sense of the Gospel texts with the sense taught by our theologians concerning the infallible knowledge of Jesus Christ." The Pope evidently does not concur with Hans Küng that the texts in question "defy any attempt to render them innocuous." Nor did the Biblical Commission condone the hypothesis of "immediate expectation" in its 1915 commentary on the Parousia.

One thing at least is clear from the start: this is no minor or

---

3. Matt. 10:23
4. *On Being a Christian* (Doubleday, 1976), p. 216.

secondary issue. The stand we take on this question will inevitably reflect—and in turn condition—our very conception of Christianity. Take the case of Hans Küng himself: no sooner does he expound his thesis of "immediate expectation" than he goes on to label the moral and spiritual teaching of Jesus as an "interim ethics," shaped by a—mistaken!—belief in an imminent end of the world. "This belief alone," he maintains, "can explain his extraordinarily urgent sayings about not being concerned for the security of life, food and clothing, about prayer being heard, the faith that moves mountains, the decision that cannot be deferred, the imagery of the great banquet and even of the Our Father and the Beatitudes."[5] Now to be sure, we absolutely disagree with each of these claims—except for one: whosoever, namely, has convinced himself that Jesus expected the Kingdom to be established within a few years is bound to have fallacious views on all these other issues as well.

~

It may be worth noting that under the Küngian auspices it becomes virtually inconceivable that Jesus should have instituted a Church charged with a worldwide mission: someone who imagines the world is about to end could hardly have proclaimed "*upon this rock I will found my church*"![6] Nor need we be surprised that those who subscribe to Küngian "imminence" should perceive the Church as instituted largely by so-called "primitive faith communities": a man-made structure thus, which may be refashioned at will.

But getting back to the eschatological logia: the first thing to note is that not all the sayings of Jesus, by any means, accord with the tenet of eschatological imminence: some apparently do, as we have seen, but others do not. To consider just one example: when Mary Magdalene anoints the feet of Jesus with a precious unguent—and Judas Iscariot objects that "this ointment could have been sold for much and given to the poor"—Jesus replies:

5. Op. cit., p. 217.
6. Matt. 16:18

"Amen I say unto you, wheresoever this gospel shall be preached in the whole world, there shall also this, that this woman has done, be told as a memorial of her."[7] Clearly, this passage, in its two parts, conveys a sense of vast duration: a time will come, after the present generation shall have long passed away, Jesus appears to say, when what this woman has done will be recalled throughout the world. The contrast between this prophecy and what Jesus proclaims in the "imminence logia" is sharp in the extreme: on the one hand the promise that this woman's deed will be commemorated by generations yet unborn, and on the other the prediction of the Parousia ere *"this generation"* shall have passed away. On the one hand the declaration that the gospel shall be preached *"in the whole world,"* and on the other the assurance to the disciples that they *"shall not have gone over the cities of Israel till the Son of man be come"*!

What, then, shall we believe? Must we join the "form critics" in their speculations regarding historical "strata" in the sacred texts? Are we obliged thus to depart from Catholic orthodoxy by distinguishing between original logia and what the *periti* perceive as post-apostolic additions? Above all, must we accept what Hans Küng—along with a coterie of like-minded pundits—opine regarding their "historical Jesus"? The exegetical problem posed by those seemingly irreconcilable logia is doubtless perplexing; but might it not, in the end, be these very so-called contradictions that will lead to its resolution? What in truth stands at issue in Christian eschatology, I contend, is a transcendence of "history"— of *time* itself—so radical that its exposition cannot but present a semblance of paradox. The contrast between the Christian and the neo-Gnostic Eschaton proves in fact to be categorical to the point of rendering these respective Termini antipodal—as antipodal ultimately as Heaven is to Hell.

～

To begin with, it behooves us to reflect upon the eschatological sermon of Mark 13 and Matthew 24-25. It is vital to observe that,

---

7. Matt. 16:13

in both Gospels, the discourse is preceded by an unequivocally historical prediction: Jesus foretells the destruction of the temple at Jerusalem. And it is in reference to this prophesy that Peter, James, John and Andrew ask Him: *"Tell us, when shall these things be? and what shall be the sign when all these things shall be fulfilled?"*[8] It is to be noted that in the parallel passage in Matthew the question reads: *"Tell us, when shall these things be? and what shall be the signs of thy coming, and of the end of the world?"*[9] It appears that in the mind of the disciples the two events—the destruction of the temple and the end of the world—are closely linked. Jesus himself, moreover, seems to connect the two to the point that it is impossible at times to tell whether He is referring to the catastrophe of 70 AD or to the Parousia. It almost seems the reference is to both at once: that these—to us exceedingly disparate—events have been telescoped to the point of fusion.

For contemporary exegetes this typically constitutes not only an enigma, but an irresolvable paradox. Yet in truth it is not; for as John L. McKenzie points out:

> One must understand that in much biblical thinking, both in the Old Testament and in the New, history and eschatology are merged in a way that is alien to modern thought... A blurring of perspective in the consideration of the fall of Jerusalem is native to it. Jesus did not depart from biblical language; and the precedents for such language in the prophetic literature were numerous enough to preclude that type of understanding that would identify the fall of Jerusalem with the beginning of the end catastrophe.[10]

To be sure, the two are by no means the same; yet the one can be seen in the other: the end catastrophe in the fall of Jerusalem. And this is not simply a matter of historical imagination, as one might think: the nexus proves to be neither subjective nor adventitious, but objectively real. The fact is that the terminal catastrophe casts

---

8. Mark 13:4
9. Matt. 24:3
10. *The Jerome Biblical Commentary* (Prentice Hall, 1968), 43:164.

its shadow, as it were, into the historical present, and half reveals itself in spectacles of stark terror. As McKenzie goes on to observe: "In particular events, such as the fall of Jerusalem, the judgment seems to break into history." And it is in this sense that the end catastrophe was indeed "at hand" when Jesus preached in Palestine, even as it may be "at hand" for us today.

Yet it is not actually the chronology of the terminal catastrophe that matters primarily, but the fact, rather, that it *will come.* From a spiritual point of vantage, this is what counts: *the End will come.* Now Jesus tells us many things about the coming of the End. He tells us, for example, that it will come suddenly, like lightning,[11] and unexpectedly, like a thief in the night;[12] but He does not enlighten us as to when it will come: *"But of that day and hour knoweth no man; no, not the angels of heaven, but my Father only."*[13] No use wracking one's brains: what alone we can know—and what alone we *need* to know—is that this *day and hour* surely *will come.* And Jesus taught us to be mindful of that day: to prepare for it, and to conform our lives, here and now, to that Kingdom which shall burst upon the world on that Day when *"the sun shall be darkened and the moon shall not give her light; and the stars of heaven shall fall, and the powers which are in heaven shall be shaken."*[14]

It has at times been suggested that the prospect of apocalyptic imminence has a commendable effect as an incentive to arouse and prod us into spiritual activity. As Joseph T. O'Connor has put it: "Oracles about future events were indeed delivered, but delivered so as to produce a change of heart in the here and now; in short, predictions about the future were motivational, not primarily cognitive factors. For that reason, there is a marked tendency in prophetic statements to *compress time sequence.* What is in fact very remote is presented as being near at hand or in the proximate future."[15] The point is well taken, and there is no reason to doubt that such is indeed the case. Yet this is by no means the whole

11. Matt. 24:7
12. Matt. 24:43
13. Matt. 24:36
14. Mark 13:24-5
15. *The Father's Son* (St. Paul Editions, 1984), pp. 125-6.

story: for as we have come to understand, *what to us appears remote may indeed be near at hand to the eye of a prophet.*[16] No longer, then, need we suppose that our "normal" experience of time is absolute; for as the Psalmist apprises us, "*a thousand years in thy sight are but as yesterday when it is past, and as a watch in the night.*"[17] So too St. Peter declares the same when he prophesies that "*in the last days*" there shall come scoffers, saying: "*Where is the promise of his coming? for since the fathers fell asleep, all things continue as they were from the beginning of creation.*"[18] "*But beloved,*" the Apostle concludes: "*Be not ignorant of this one thing, that one day is with the Lord as a thousand years, and a thousand years as one day.*"[19]

The reason, thus, why the New Testament references to the Parousia may appear equivocal to the point of contradiction lies deeper, by far, than Hans Küng and his ilk are able to comprehend. If only one reads these scriptural texts with so much as a modicum of metaphysical discernment, one comes to realize soon enough that the apparent contradictions are in fact indicative of a vision that transgresses the bounds of time. Strange as it might strike contemporary ears, it proves to be our *periti*—and *not* Jesus—who are duped.

∾

But let us go on. Up till now we have spoken in terms of "near" and "far," and thus in inherently quantitative terms. Yet if one looks closely at the pertinent logia, one sees that much more stands at issue. Consider for instance the logion: "*Amen, I say unto you, the hour is coming, and now is, when the dead shall hear the voice of God; and they that hear shall live.*"[20] Think of it: an hour yet to come "*now is*"! What saves this, however, from reducing to a flat contradiction is the fact that two radically different modes of knowing are in play. We need to ask, in the first place, what specific

16. We have dealt with this issue in chapter 10.
17. Psalm 90:4
18. 2 Pet. 3:3-4
19. Ibid., 3:8
20. John 5:25

event is signified by that "hour" to which Christ alludes: might it be the Resurrection? Or God's Judgment upon the world? Or perhaps the Second Coming of the Lord? One is forced finally to admit that it is in truth all of these—and yet it is one definitive and ultimate Event. With the dawn of the first Easter, namely, time has been "fulfilled": the Kingdom of God has, in a sense, "broken in upon the world." The Resurrection marks the advent of "the Eighth Day": the Day of the New Creation, which shall have no end. As Scripture declares, Christ has triumphed *"once for all."*[21]

"The hour is coming, *and now is*": it *now is* in Christ. But as for us, who ourselves are not yet fully "in Christ," that *hour* has not yet struck: for us that hour is still *coming. The Lord is coming*: down the centuries He comes, and *"He which testifies these things saith, Surely I come quickly. Amen. Even so, come, Lord Jesus."*[22]

Since the Resurrection history has been moving towards a Destination which already exists. As F. X. Durrwell observes in somewhat "iconic" terms: "Since Easter, human time has been advancing towards an event of the past, the Resurrection of Christ, and it will only reach it at the end of history. Our Lord has passed out of time and become its Center because He is its fullness, its Master because He is its End."[23] Admittedly the image of time "advancing towards an event of the past" may be but a figure of speech; yet it can serve, even so, to jolt us into the recognition that the Resurrection of Jesus Christ has broken the bounds of history, and has in truth "passed out of time and become its Center." If in fact it had *not* done so, it could not have the universal and enduring significance which in truth it has: it would then be indeed a "past event." So far, however, from conceding that claim, Christianity insists that our hope of eternal life is predicated upon the possibility of *sharing in that very Event!*

It is a fact that the authentically Christian life—both here and hereafter—is centered upon the Death and Resurrection of Christ. Yet, timebound as we normally are, we find it hard to conceive

21. Heb. 10:10
22. Rev. 22:20
23. *The Resurrection: A Biblical Study* (Sheed and Ward, 1960), p. 263.

of these seemingly remote and successive "events" in other than exclusively historical terms. Only at rare moments do we sense that the mystery of Easter actually resides—not in the distant past—but in the immediate present: all too rarely does it dawn upon us that "*now is the accepted time*," that "*now*" indeed is "*the day of salvation*."[24]

∾

In conformity with our time-bound outlook, theology is obliged to distinguish between the Resurrection and the Second Coming of Christ. One thus speaks typically as if these were two distinct and successive events: the one in the distant past, the other in the indefinite future. And yet it has occasionally been recognized that the Resurrection of Christ "*is* in fact the final coming, the only coming, upon the clouds, at the right hand of the Power."[25] One and the same Event, one might say, is perceived variously from different points of vantage. As Durrwell goes on to explain:

> Seen in Christ, the reality of the *Parousia* is contained in the Resurrection by being actually present and wholly realized; the glorification, the destruction of death,[26] the final resurrection, the judgment, the subjection of the powers—all these are realized from that time onwards. Seen from the point of view of mankind, the *Parousia* is gradually being brought about until it blazes forth on the last Day; but this is simply the realization in humanity of God's judgment which is already in the world, and of the one Resurrection of Christ... Time, which for us flows continuously between Christ's Resurrection and the *Parousia*, is as it were contracted in Christ's exaltation; for us on earth it shows one by one the effects of the *Parousia* of Christ, which will eventually be revealed as a whole.[27]

Having noted that the authentically Christian life is centered

24. 2 Cor. 6:2
25. Op. cit., p. 268.
26. 2 Tim. 1:10
27. Op. cit., pp. 268 & 255.

upon the Death and Resurrection of Christ, it should by now be clear that this is not simply a matter of sentiment, but of "ontological participation" as one might say. To be sure, the Christian does of course recall—with love and reverence—the salient events in the life of Jesus, even as he looks forward to the Second Coming of the Lord. Yet the fact remains that his connection with these past and future events is by way of the present: that it hinges upon a certain "indwelling" of the Risen Lord "who has passed out of our time and become its center." Whether the Christian realizes it or not, here and now already he lives a double life: for he lives not only on the temporal Circumference, but also already within the Christic Center.

It may be true that, for most of us, the focal point of our life remains yet close to that Circumference: after all, we are still predominantly earthbound creatures. But the balance may shift in an instant: it all depends upon the intensity of our spiritual realization. The more closely united to Christ we become, the more radically the world will cease in us: "the end of the world is realized in the believer to the extent that he shares in the Resurrection."[28] As Christ proclaims, the Parousia is "at hand" and can burst upon us at any moment. It happened to St. Stephen, the first Christian martyr,[29] and after him to countless saints. From the earliest Christian times to the present day there have been those *which shall not taste death till they see the Son of man coming in his kingdom.*

∿

Having so far spoken in predominantly "iconic" terms, it behooves us finally to consider the eschatological issue at hand from a metaphysical point of vantage. Let it be clearly understood, in the first place, that the Christian Eschaton—known as the Kingdom of God—transcends the cosmos *in its entirety*. Not only, thus, is that realm untouched by the bound of time, but it transcends even what we have termed the aeviternal plane. That Kingdom is therefore

---

28. Ibid., p. 269.
29. Acts 7:55-6

beyond the reach even of Platonist philosophy. And let us note that this is why St. Thomas Aquinas distinguishes categorically between "eternity," properly so called, and "aeviternity," which he characterizes by the fact that "time may be added to it." The aeviternal realm constitutes namely what we have termed the Center of the tripartite cosmos, whereas the Kingdom of God, being *eternal* in the strict sense, transcends *ipso facto* the cosmos in its entirety. As Christ declared to Pontius Pilate: "*My kingdom is not of this world.*"

It needs now to be understood that, so far from constituting an event—something that transpires "in time"—the Parousia is rather to be conceived as that which *terminates* time itself: puts an end to it once and for all. And that is the reason why the Parousia can "manifest," as we have seen, in events ranging over an interval of time: from the Resurrection, that is, right up to that "*day and hour*" which "*no man knoweth.*" Let us note, finally, in light of what we have said *apropos* of the Edenic state, that the Parousia is not visible even to the eye of a prophet by virtue of the fact that it transcends the aeviternal realm itself.[30]

The Hans Küngs of this world are thus egregiously mistaken when they conceive of the Parousia as "an event," as something, namely, that happens in time: as we have seen, the very opposite holds true. And whereas the final termination will come on that fateful day "*no man knoweth,*" the Parousia is able—by the very fact of not being subject to temporal bounds—to manifest itself in an indefinite number of "apocalyptic" events: for instance in the catastrophe of 70 AD. Ever since the Resurrection of Christ the Parousia has in fact been "near at hand" for the Christian, so much so that many a saint has witnessed that Second Coming at the moment of death.

A rather different point is made in the logion of Matthew 24:34, where Christ declares: "*Amen I say unto you, this generation shall not pass till all these things are fulfilled*"—for it is to be understood that the fulfillment in question was achieved precisely through the Death and Resurrection of Christ. And so also, when

30. See ch. 10.

Jesus declares in the Gospel of St. Mark that *"the time is fulfilled and the kingdom of God is at hand,"* He did so in anticipation of His redemptive Act.

And this, at last, resolves the enigma which has induced the likes of Hans Küng to attribute "human error" to the Savior Himself. So far from being erroneous, it turns out the logia in question are expressive of a supra-human truth: the very *truth* in fact, which in the end—*Deo volente*—*"shall set us free."*

≈

Let us then reflect, finally, upon that which has rendered us "bound." As noted in chapter 10, that mythical "event" has been identified in the Judeo-Christian tradition as the Fall of Adam, and his subsequent Expulsion from the Garden of Eden. With the understanding that this is not, strictly speaking, history, but rather "pre-history," one can say that this is the pivotal event which has caused mankind to be "cut off" as it were from the aeviternal plane, and thereby reduced to the status of a *psychikos anthropos.*

I would like, first of all, to point out, based upon the doctrine of the Parousia, that the End to which Christ calls us is incomparably higher than that Edenic state: as has ever been recognized by the wise, the micro- and macrocosm combined are as nothing compared to their transcendent Cause. It thus becomes apparent that the Fall of Adam was indeed a *felix culpa*—a "happy fault"— inasmuch as it has in a sense resulted in opening the way to the supreme Eschaton. What Christ terms "the Kingdom of God" pertains no longer to the cosmos—to the Creation—but to God, the Creator Himself. The Gospel or Good News of Christianity takes the form, thus, of an invitation so literally *wonderful* as to be barely conceivable: the fact is that *each of us has been invited—"by name" as it were—to what is termed a Wedding Feast*—yes, *by none other than God Himself.* Believe it or not—accept it or not—that is the choice each of us shall have to make.

Such, in all brevity, is the Invitation definitive of the Christian religion. I emphasize this point because it has so often been implicitly denied by those who would subsume Christianity under the rubric of a *religio perennis* exemplified, first of all, by the oldest religious tradition in the world, which appears to be the Vedic. Yet with all love and reverence for Hindu religion, it needs to be understood that—search where you will—such an Invitation is nowhere to be found within the Vedic domain.[31] One can always, of course, explain away the Offer, be it as an exoteric simplification—a mere metaphor—or an outright beguilement: the question can be argued endlessly. What is by no means debatable, on the other hand, is the uniqueness of the Christian Calling: we know precisely *when* and *where* it was first proclaimed, and by *Whom*.

---

31. I have touched upon that issue in *Physics and Vertical Causation* (Angelico Press, 2019), pp. 115-22.

13

DOES PHYSICS ADMIT
A TELEOLOGY?

HAVING AT LAST discovered that the tripartite cosmos does not stand alone—that the Christian Eschaton transcends even its aeviternity—we have arrived at a metaphysical affirmation of God. And so the question arises whether "the existence of God" has implications of a cosmological kind: whether it tells us something about the cosmos itself. I would like now to show that most assuredly it does, and that in fact it answers a question which cannot otherwise be resolved: i.e., *does physics admit a teleology?*

Given that God constitutes indeed "the first Cause and last End of all things," it would appear that the reach of theology extends in principle to the cosmos as well, which is to say that there are truths concerning the cosmos that can be elicited only from a theological point of vantage. A case in point, I claim, is the fact that physics *does in truth admit a teleology.*

My first point is that physics as such cannot enlighten us on that score, for the obvious reason that the very notion of a *telos*—of an end or objective—is distinctly beyond its ken. And this holds true for both classical as well as quantum physics: as Whitehead pointed out a century ago, what physics portrays is no more than "the hurrying of material, endlessly, meaninglessly" —which is to say that if the cosmos were in truth what the physicist takes it to be, there would be no *telos* at all.

~

We commence our inquiry then with the formal recognition that the cosmos does not stand alone: that it is neither self-caused nor self-sufficient, but is both brought into existence and "held in being" by a metacosmic Reality, which though variously conceived and designated, we will refer to—in keeping with the Judeo-Christian tradition—by the venerable designation *"God."* And we propose now to take our cue from Meister Eckhart, the medieval master who appears to have probed the cosmic implications of the ultimate metaphysics more profoundly than most theologians. Let us begin, then, by contrasting his *Weltanschauung* with the physicist's perception of Nature as "a dull affair: merely the hurrying of material, endlessly, meaninglessly." "Not so" responds Eckhart in effect; on the contrary: *"All creatures are by nature endeavoring to be like God"*! The contrast is as profound as it is razor-sharp—and the Meister goes on:

> *The heavens would not revolve unless they followed on the track of God or of his likeness. If God were not in all things, Nature would stop dead, not working and not wanting; for whether thou like it or no, whether thou know it or not, Nature is seeking, though obscurely, and tending towards God.*[1]

It should be noted, first of all, that the idea of Nature *"seeking, though obscurely, and tending towards God"* is by no means unprecedented in Christian tradition: has not St. Paul presented us with the reverse side of the same doctrine when he declares that *"the whole creation groaneth and travaileth in pain together until now"*?[2] Whether or not, therefore, there is to be a "return" of the universe to its Source—an *apocatastasis*—it is that End, that supreme *telos* itself, that impels the cosmos to pursue its trajectory.

But clearly: given that the *modus operandi* of physics deals exclusively with the quantitative dimensions of the cosmos, none of this has meaning for a physicist as such. Yet even so, if Nature indeed

---

1. *Meister Eckhart.* Trans. C. de B. Evans (London: Watkins, 1925), vol. I, p. 115.
2. Rom. 9:22

"*groaneth and travaileth*" seeking "rest in God"—if that is more than simply a pious metaphor—how could this fact *not* impact the actual laws of physics? There must consequently be a connection, a nexus of some kind, between the aforesaid *telos* and the actual equations of motion—whether that nexus be comprehensible to the physicist or not. And what is more: that "scientifically invisible" connection should be discernible from a metaphysical point of vantage—that is what we propose now to verify.

The difficulty, of course, is that the laws of physics, given as they are in mathematical terms, bear no visible reference to a *telos*—let alone to a Pauline "travail" or Eckhartian "rejoicing"! It is hard to imagine, therefore, how one could possibly discover in these formulae a spoor of the *telos*. And yet, from the aforesaid metaphysical point of vantage one would expect some feature, some "mark" of the impelling End to manifest in the mathematical formalism of physics itself.

Let us return, then, to Meister Eckhart and fix our attention on the phrase "*seeking, though obscurely, and tending towards God.*" Never mind that this notion strikes the contemporary scientific mind as infantile: our job is to see whether or not it connects with the equations of physics. And the first thing that strikes one in that regard is that, by implication, Nature *moves*—not for the sake of motion—but precisely for the sake of *rest*. Hence it moves *sparingly* in some appropriate sense. In fact, of all the ways to get from a state A to a subsequent state B, Nature will choose a path which, according to some metric, requires what may be aptly termed the least "expenditure of motion." We have put the phrase under quotes because that metric remains to be identified. Yet, by now, the concept in question stares us in the face: it can in fact be none other than what physicists term "action," which is defined not simply for a particle, but for a physical system as a whole, subject to given forces. And there exists indeed a law of the kind we are seeking, which for centuries has been known as *the principle of least action*.

Dating back to the early days of Newtonian mechanics, it is associated chiefly with the names of Euler, D'Alembert, Lagrange, and Hamilton, and has been formulated in various ways. As "the principle of least action" properly so called—that is to say, in its

Lagrangian formulation—it affirms that in moving from a state A to a state B, a physical system will choose the path for which a certain integral—definitive of "action"—attains a minimum (or at least a "stationary") value. Let me remind the mathematically informed reader that what stands in question is the integral over a path from A to B in the configuration space of the Lagrangian function L, defined as the kinetic minus the potential energy of the system. The fact is that the equations of motion for classical physics can be obtained from the given integral via the calculus of variations—which is to say that, of all the possible paths from A to B, Nature chooses the path of "least action." Certainly there are other ways of deriving the classical equations; yet it appears that the principle of least action takes precedence over these: for one thing, among all these ways—all workable, all useful—only the principle of least action extends to the quantum world.[3]

What relates that "principle of least action" to the Eckhartian tenet is the fact that, inasmuch as the Reality underlying the cosmos transcends both space and time—and therefore undergoes no "action" at all—the best likeness or approximation to the metacosmic state a trajectory in a configuration space can attain is specified by the principle of least action. So far, therefore, from being void of implications for physics as such, the Eckhartian stipulation leads directly to what has long been deemed the most elegant derivation of classical dynamics, which extends into the quantum realm and proves in fact to be the master key to the laws of physics at large. In addition to its purely physical significance, the principle of least action carries thus a *teleological* sense undetectable on the basis of physics *per se*. Little does the physics community surmise that this celebrated principle accords in truth with the "infantile" notion that the universe at large is indeed *"seeking, though obscurely, and tending towards God"*!

3. See Natalie Wolchover, "A Different Kind of Theory of Everything," *The New Yorker*, 19 February 2019.

14

SCIENCE, SCIENTISM, AND SPIRITUALITY

HAVING BROUGHT to light the tripartite ontology of the integral cosmos as well as the radical transcendence of the Christian Eschaton, we have provided a basis for the consideration of the ultimate question—touching upon the supreme desideratum—which cannot but pertain to the province of religion. I propose now to reflect upon the impact of both science and scientism upon that domain, and will do so from a Christian and indeed a Catholic point of vantage, which conforms after all to the teachings of Christ as taught by the Apostles.

Let me begin with the observation that authentic science *per se*—dissociated thus from its nimbus of scientistic beliefs—neither is nor *can be* in conflict with Christianity, for the simple reason that the very origin of truth resides evidently in God, who is after all the first Cause and last End of all that exists. The point can in fact be made that insofar as science is a quest of truth, it is inherently a *Christian* enterprise in light of the Christic logion "*I am the truth.*" The notion, therefore, that there is an inherent conflict or incompatibility between science and Christian belief is misconceived; and as we have come to understand, is actually based upon a failure to distinguish between science, properly so called, and scientism.

What needs to be grasped is that scientism is itself a kind of religion, which as we have seen, can in fact be identified as the counter-religion to Christianity. And this recognition, I say, proves to be the key to the understanding of the crisis presently playing out before our eyes. It must not be thought, however, that the problem is simply a matter of "ideology" in the current sense of "values" as distinguished from "facts." To think thus is to have already succumbed to the counter-religion! It is to forget that the current dichotomy of "facts versus values" is itself part of the neo-Gnostic strategy, giving its pundits free reign to parade all kinds of anti-Christian myths as truths of "science," which is supposedly "value-free" and consequently "objective."

This brings us to our major point: some time following the Galileo confrontation, after the new physics and its associated sciences had gained a degree of status and influence over Western society, it apparently became the consensus within the Church to disengage from cosmological issues, to turn these matters over to the scientific community to deal with as they see fit, and not to challenge the scientists even when they contradict—overtly or by implication—what in days gone by the Church had taught on the issues in question. What does it matter, her spokesmen seemed to say, what we believe regarding the position of the Earth in the cosmos or the composition of corporeal entities—or even on the issue of evolution—so long as we remain orthodox in matters of religion. In response let me first of all state apodictically that it *does* matter: that scientism namely, in any of its modes, proves to be not only deleterious, but in fact toxic to the spiritual life. We must not forget that religion—so long as it has not degenerated into convention or mere sentimentality—demands the whole man: *holiness* and *wholeness* are ultimately inseparable. Does not "*the first and greatest*" commandment enjoin that one must love "*the Lord thy God*" not only "*with all thy heart and soul,*" but also "*with all thy mind*"! What we think about the world—our *Weltanschauung*—cannot with impunity be excluded from the sphere of religion; as St. Thomas Aquinas states in *Summa Contra Gentiles*:

It is absolutely false to maintain, with reference to the truths of our faith, that what we believe regarding the creation is of no consequence, so long as one has an exact conception of God; because *an error regarding the nature of creation always gives rise to a false idea concerning God*.[1]

1. Bk. II, ch. 3; italics mine.

One might add that the history of Western civilization since the Enlightenment amply confirms this Thomistic principle: from the deism of Voltaire and the atheism of Laplace right up to the "science-fiction theology" of Pierre Teilhard de Chardin[2] one beholds the spectacle of scientistic errors begetting spurious and indeed "lethal" theologies.

Clearly, what we hold to be true regarding matters of science does affect our theological beliefs and most assuredly impacts our spiritual life. Moreover—with due allowance for what might be termed "invincible ignorance"—it cannot be denied that we are in some measure responsible for what we hold to be true in this supposedly "secular" domain of inquiry: *"with all thy mind"*—these four words alone should suffice to dispel all doubt in that regard. I would further contend that religion goes astray the moment it capitulates to scientific authority when it comes to "what we believe regarding the creation." It appears that the ongoing de-Christianization of Western society has a great deal to do with the fact that, ever since the Enlightenment, our cosmology has been abandoned to the mercy of the scientists—to the neo-Gnostic gurus ultimately!

Yet the matter does not end with a factitious cosmology: for it is the ineluctable tendency of science to absolutize the cosmos, which comes thus to usurp the place of God. As Theodore Roszak, for one, has noted with admirable clarity: "Science is our religion, because we cannot, most of us, with any living conviction see around it."[3] And one might add that Oskar Milosz has made much the same point even more sharply: "Unless a man's concept of the physical universe accords with reality," he writes, "his spiritual life will be crippled at its roots, with devastating consequences for every other aspect of his life."[4] Would that these authors were

---

2. I have expressed my views regarding Teilhard de Chardin in *Theistic Evolution: The Teilhardian Heresy* (Angelico Press, 2012).

3. *Where the Wasteland Ends* (Doubleday, 1973), p. 124.

4. Cited by Seyyed Hossein Nasr in *Religion and the Order of Nature* (Oxford University Press, 1996), p. 153. Concerning Oskar Milosz, see Philip Sherrard, *Human Image: World Image* (Golgonooza Press, 1992), pp. 131-46.

studied and reflected upon in our seminaries by students and professors alike!

~

The capitulation of the Church to science in matters of cosmology has rendered her vulnerable to the scientistic worldview; and as regards the implications of this fact for the life of the Church, let me quote the French philosopher Jean Borella:

> The truth is that the Catholic Church has been confronted by the most formidable problem a religion can encounter: the scientistic disappearance (*disparition scientifique*) of the universe of symbolic forms which enable it to express and manifest itself, that is to say, which permit it to exist.

And he goes on to say:

> That destruction has been effected by Galilean physics . . . because it reduces bodies, material substance, to the purely geometric, thus making it at one stroke impossible (or devoid of meaning) that the world can serve as a medium for the manifestation of God. The theophanic capacity of the world is denied.[5]

It should be noted that Borella is implicitly referring to what I term the reduction of the corporeal to the physical: "*le problème le plus redoubtable qu'une religion puisse rencontrer*"—"the most formidable problem a religion can encounter"—he calls it. What he terms "a reduction to the purely geometric" moreover refers first of all to Cartesian bifurcation, and ultimately to the Einsteinian project discussed in chapter 3. It is this "reduction to the purely geometric" that obliterates "the theophanic capacity of our world."

It is of course to be understood that the "symbolic forms" to which Borella alludes are not—as some might imagine—subjective images or "ideas" which, in bygone days, mankind had naïvely

---

5. *Le sens du surnaturel* (Éditions Ad Solem, 1996), p. 74.

projected upon the external universe—until, that is, Science arrived upon the scene to enlighten us. The very opposite of this officially mandated scenario is in fact the case: the "symbolic forms" to which Borella refers are objectively real and comprise in truth the very foundation of the universe. We may conceive of them as "forms" in the Aristotelian and Scholastic sense, or Platonically as eternal archetypes reflected upon the plane of cosmic existence. In either case they constitute the very essence of corporeal being: remove these "symbolic forms" and the cosmos ceases to exist, for it is they that anchor the cosmos to God.

The crucial fact is that the scientistic denial of the corporeal entails a negation of the substantial forms and essences which make up the order of being, along with the sensible qualities which manifest these forms and essences. The scientistically conditioned mind has thus become *de facto* incapable of recognizing what Borella terms "the universe of symbolic forms," and it is in this sense that "the theophanic capacity of the universe" has disappeared.

It is hardly surprising, then, that the consequences of this breach prove to be tragic in the extreme. In his denial of essences—which in truth are comparable to a light penetrating into our world from a higher sphere—scientistic man has jettisoned the very basis of the spiritual life: has in effect obliterated what "enables the Church to express and manifest itself," and in so doing, "permits it to exist."

The refutation of scientistic belief is therefore a *sine qua non* for the restoration and indeed the very survival of the Church on Earth: no wonder its visible manifestation hardly resembles the Church anymore. "Progressive" spokesmen will of course respond by holding forth on what they love to characterize as a "return to the Middle Ages"—yet the fact remains that a partial such "return" in the form of an expulsion of scientistic heresy is today, for the Church at large, a matter not only of urgent necessity, but actually of *survival*. The stakes could not be higher! As I have noted earlier in reference to fundamental physics,[6] so too in the ecclesiastic

6. See ch. 4.

sphere it appears that we are fast approaching a critical point, a "singularity" at which the present trajectory cannot but come to an end. What happens beyond this ominous point—that, I surmise, is a question to which Scripture alone holds the key.

~

It will be enlightening, at this juncture, to recall what St. Paul reveals to us regarding "the theophanic capacity of the world" in his Epistle to the Romans:

> *For the invisible things of Him from the creation of the world are clearly seen, being understood by the things that are made, even His eternal power and Godhead.*

To which he adds:

> *So they are without excuse: Because that, when they knew God, they glorified Him not as God, neither were they thankful; but became vain in their imaginations, and their foolish heart was darkened. Professing themselves wise, they became fools.*[7]

I need hardly point out the striking relevance of these words to all that we have touched upon! "*Professing themselves wise, they became fools*": is this not in fact a perfect characterization of our scientistic predicament? The "*things that are made*" can evidently be identified as *corporeal* entities, the objects mankind normally perceives in the waking state. And what about "*the invisible things of Him*": are these not precisely what we have referred to as eternal essences, ideas, or archetypes? So long as our heart has not been "*darkened*," the "*things that are made*" will awaken in us an intellectual perception—a "recollection" as Plato says—of the eternal things they reflect or embody.

St. Paul alludes to a time or state when man "*knew God*," a reference, first of all, to the condition of Adam before the Fall,

---

7. Rom. 1:20-2

when human nature was as yet undefiled by Original Sin.[8] One needs however to realize that the Fall of Adam has been repeated on a lesser scale down through the ages, in an unending series of betrayals, large and small. Even today, at this late stage of human history, we are yet, each of us, endowed with a certain "*knowledge of God*" to which we can freely respond. And that is precisely why we, too, are "*without excuse*," and why, to some degree at least, we are responsible for what we take the cosmos to be. Everyone perceives the world ultimately in keeping with his spiritual state: the "*pure in heart*" perceive it without fail as a theophany—and as for the rest of us, "*whose foolish hearts are darkened*," the theophanic capacity of the universe is reduced in proportion to that very darkening.

Not only, however, does our spiritual state affect the way we perceive the external world, but it is likewise to be noted that, conversely, the way we perceive the world—literally our *world-view*—reacts invariably upon that spiritual state. What we take the universe to be—how we look upon the world—has a profound effect upon our spiritual life. The decisive point I wish to convey is that the scientistic *Weltanschauung* takes us in principle to a state which may actually be characterized as *subhuman*. Happily, however, human nature as such prevents us normally from embracing wholeheartedly what the pundits of the flatland proclaim: to be fully cut off from "the theophanic capacity of the universe" is after all, for us humans, unbearable.[9] Whether we know it or not—and whatever our scientistic *periti* may say—we stand in quotidian need of that "theophanic capacity" even as plants demand light and animals require food.

The fact is that man and cosmos are inseparably linked. Let us recall that according to the traditional wisdom of mankind, the *anthropos* actually constitutes a *microcosm*: a "cosmos in miniature," a fact which entails that even as man is tripartite—consisting of *corpus*, *anima*, and *spiritus*—so too is the cosmos at large.

---

8. On this subject I refer back to chapter 10.

9. This clearly holds true for the pundits of scientistic materialism as well. Who can actually imagine a man, say, who regards his wife or child as an ensemble of quantum particles? As my esteemed friend Olavo de Carvalho says again and again: "They are lying!"

It is hardly surprising, then, that as our conception of the cosmos flattens, so does our conception of man. Contemporary man does not have a high estimate of himself! Is it any wonder then when his goals and ideals follow suit? How can someone conditioned to regard the subcorporeal as the real—and hardly believes any longer in the redness of a rose—be open to "the theophanic capacity" of the universe? We must remember that this theophanic capacity resides neither in the corporeal nor even in the intermediary plane, but precisely in their trichotomous integrality: the very wholeness post-Galilean science has been at pains to obliterate by a systematic process of "atomization."

It appears that "what we believe regarding the creation" does matter, and that it was in truth a fatal blunder to abandon cosmology to the mercy of the scientist. Is it any wonder that "the Catholic Church has been confronted by the most formidable problem a religion can encounter"? The restitution of the Church—which we are assured will come in the ripeness of time—demands that we break through the barrier of scientistic belief and return to ontological orthodoxy. And this rectification constitutes not merely a desideratum, but an absolute necessity, as indeed the neo-Gnostic nature of the scientistic deviation makes clear.

INDEX OF NAMES

Aquinas, St. Thomas xix, xxiii, xxvi, 1, 17, 20 fn., 63, 69, 113 fn., 116, 131, 165, 174

Aristotle xvii, xix, 1, 27, 30, 35, 57 fn., 66, 76, 79, 80, 83, 109, 117 fn., 128

Augustine, St. 108 fn., 138 fn., 146

Bacon, Francis 121

Balthasar, Hans Urs von 150

Bennett, M. R. 60-2, 64-5

Bohm, David xxii, 18, 71-6, 78-9, 81-3, 129 fn.

Bohr, Niels xvii, xxii, 37, 55, 95

Borella, Jean xvii, xviii fn., xxvi, 28, 29-34, 36, 85 fn., 89, 94, 176-7

Bortoft, Henri 128, 129 fn.

Brentano, Clemens 134

Burtt, Edwin A. 100

Coomaraswamy, Ananda 28, 100

Crick, Francis 60, 61, 126

D'Alembert, Jean le Rond 171

Dante Alighieri 32 fn., 112, 113

Darwin, Charles xxv, 29 fn., 54, 102-6, 108-9, 115, 122-3, 149-50

Davies, Paul 48

de Broglie, Louis 74

de Carvalho, Olavo 179 fn.

de Fiore, Joachim 146

de Lubicz, Schwaller ix

Dembski, William 12, 18, 90 fn., 92, 102, 115

Democritus 5 fn., 57 fn., 59, 79, 109 fn.

Descartes, René x-xii, xviii, xix, 3, 5-7, 40, 57, 59, 77, 96 fn.

Durrwell, F. X. 162, 163

Eddington, Arthur 22, 24, 25, 37

Einstein, Albert xvii, xxiv, 12, 14, 18, 23, 39-40, 52-4, 72, 106-8, 122-3, 150, 176

Ellis, George F. R. xxii fn., 51

Emmerich, Anna K. 134-5

Euler, Leonhard 171

Feynman, Richard 21, 45, 53, 55, 70

Florensky, Pavel 83

Freud, Sigmund xxv, 97 fn., 123, 140

Galilei, Galileo xix, 27, 32, 36, 40, 41, 59, 66, 174

Gibson, James J. xi, xxii, 8-9, 65-8, 77-8, 80-3, 130-1

Gödel, Kurt 93 fn.

Goethe, J. W. 2 fn., 55, 129 fn.

Gould, Stephen Jay 104 fn.

Greene, Brian 48

Guénon, René xvii

Hacker, P. M. S. 60-2, 64-5

Hamilton, William R. 171

Hawking, Stephen xvii, 48

Hayek, F. A. 99

Hegel, G. W. F. 148

Heisenberg, Werner xvii, xxii, 2 fn., 11, 22, 24, 25, 37, 47, 69, 72 fn., 73 fn., 75, 78

Heraclitus 13

Herbert, Nick 47

Hiley, B. J. 71-3
Hill, Andrew R. 118 fn.
Hossenfelder, Sabine 24 fn., 39, 45,
    47-56
Hume, David 97
Huxley, Aldous 135

Irenaeus, St. 139, 154

Johnson, Phillip 104
Jonas, Hans 139
Jung, Carl xxv, 6, 97 fn., 123, 148

Kaku, Michio 48, 53
Kant, Immanuel xix, 97
Kepler, Johannes 95
Koyré, Alexandre xviii
Kuhn, Thomas 8, 103, 104
Küng, Hans 156-8, 161, 165, 166

Lagrange, Joseph-Louis 171
Laplace, Pierre-Simon 175
Lings, Martin 149
Locke, John 97
Lord Kelvin 42, 72, 121

MacAndrew, Alec 1-12
Magus, Simon xxv, 139
Marcion 139
Martin, Malachi 114 fn.
Marx, Karl 148
McKenzie, John L. 159-60
Medawar, Peter 105
Meister Eckhart 117, 170-2
Mignot, Jean 96, 97 fn.
Milosz, Oskar 175
Moses 14 fn.

Newton, Isaac x, xviii-ix, 17, 36, 40,
    43, 47, 57, 60, 69-70, 100, 102 fn.,
    171
Nietzsche, Friedrich 35, 147, 148

O'Connor, Joseph T. 160
Olive, Keith 54

Paul, St. xxvi, 29, 34, 63 fn., 87 fn.,
    131, 132, 138, 140, 141, 156, 170,
    171, 178
Pauli, Wolfgang 44
Peterson, Gregory R. 99
Pius X, Pope St. 156
Plato xvii, xix, xxii, xxiii, 1, 27, 28-32,
    34-5, 57 fn., 80 fn., 85-7, 89, 94,
    96-7, 109, 113-4, 116, 118-9, 130,
    132, 133, 177, 178
Plotinus 141
Popper, Karl 99, 103
Pythagoras 34, 91, 96, 97 fn., 123

Raphael 35
Rieff, Philip 7
Roszak, Theodore 175
Rousseau, Jean-Jacques 140, 148
Russell, Bertrand 91 fn., 102 fn.

Schrödinger, Erwin 23, 37
Smith, Huston 144
Strassler, Matt 48

Teilhard de Chardin, Pierre xxv, 105,
    106, 108, 150-3, 175
Thomson, J. J. 43
Tully, Brent 106 fn.

Valentinus xxv, 139
Voegelin, Eric 145, 147, 148-9
Voltaire 148, 175

Wallace, William A. 10 fn.
Weinberg, Steven 24, 55
Whitehead, A. N. xviii, 57, 60, 80
    fn., 91 fn., 98, 102 fn., 169

ABOUT THE AUTHOR

WOLFGANG SMITH was born in Vienna in 1930. At age eighteen he graduated from Cornell University with majors in physics, mathematics, and philosophy. At age twenty he took his master's degree in physics at Purdue University, subsequently contributing a theoretical solution to the re-entry problem for space flight while working as an aerodynamicist at Bell Aircraft Corporation. After taking his doctorate in mathematics at Columbia University he pursued a career as professor of mathematics at M.I.T., U.C.L.A., and Oregon State University until his retirement in 1992.

Notwithstanding his professional engagement with physics and mathematics, Wolfgang Smith is at heart a philosopher in the traditional sense. Early in life he became deeply attracted to the Platonist and Neoplatonist schools, and ultimately undertook extensive sojourns in India and the Himalayan regions to contact such vestiges of ancient tradition as could still be found. One of the basic lessons he learned by way of these encounters is that there actually exist higher sciences in which man himself plays the part not merely of the observer, but of the "scientific instrument": i.e., becomes himself, as it were, the "microscope" or "telescope" by which he is enabled to access normally invisible reaches of the integral cosmos. By the same token, Smith came to recognize the stringent limitations to which our contemporary sciences are subject by virtue of their "extrinsic" *modus operandi*: the folly of presuming to fathom the depths of the universe, having barely scratched the surface in the discovery of man himself.

Following his retirement from academic life, Smith devoted himself to the publication of books missioned to correct the fallacies of contemporary scientistic belief by way of insights derived from the perennial wisdom of mankind. These works focus primarily on foundational problems in quantum theory and the no less challenging quandaries related to the problem of visual perception. The key to the overall puzzle, according to Smith, is to be found in the long lost cosmology of antiquity, which conceives of the integral cosmos as tripartite—even as man himself is traditionally viewed as a composite of *corpus*, *anima*, and *spiritus*.

The Philos-Sophia Initiative Foundation has produced a feature documentary on the life and work of Dr. Smith, *The End of Quantum Reality*, which is available on disc and digital platforms worldwide. Visit theendofquantumreality.com for more information.